# Racing Math

**CHECKERED FLAG ACTIVITIES AND PROJECTS FOR GRADES 4 - 8**

Barbara Gregorich and Christopher Jennison

Illustrated by Douglas Klauba

Good Year Books
*An Imprint of Addison-Wesley Educational Publishers, Inc.*

## Dedication

*For Jon Garfield, friend and teacher.
With special thanks to Matthew Dempsey
for suggesting the topic.*

Good Year Books are available for most basic curriculum subjects plus many enrichment areas. For more Good Year Books, contact your local bookseller or educational dealer. For a complete catalog with information about other Good Year Books, please write:

**Good Year Books
Scott Foresman - Addison Wesley
1900 East Lake Avenue
Glenview, IL 60025**

Design by Patricia Lenihan-Barbee.
Text copyright © 1998 Barbara Gregorich and Christopher Jennison.
Illustrations copyright © 1998 Good Year Books, an imprint of Addison-Wesley Educational Publishers, Inc.
All Rights Reserved.
Printed in the United States of America.

ISBN 0-673-36376-6

1 2 3 4 5 6 7 8 9 -EB- 06 05 04 03 02 01 00 99 98

Only portions of this book intended for classroom use may be reproduced without permission in writing from the publisher.

## Photo credits

Unless otherwise acknowledged, all photographs are the property of Scott Foresman - Addison Wesley. Page abbreviations are as follows:
*t* top, *c* center, *b* bottom,
*r* right, *l* left.

Front cover:
- *l* Pascal Rondeau/ Allsport
- *c* Focus on Sports
- *r* Steve Ellis/ © Indianapolis Motor Speedway

Back cover:
- *tl* Focus on Sports
- *bl* Caputo / Gamma-Liaison

Insert page 1:
- *tl* Focus on Sports
- *tr* Roger Berlwell / © Indianapolis Motor Speedway
- *cl* Steve McManus / © Indianapolis Motor Speedway
- *l* Mike Powell / Allsport
- *bl* Jim Haines / © Indianapolis Motor Speedway
- *br* Armando Franca / AP/ Wide World

Insert page 2:
- *t* Ron McQueeney / © Indianapolis Motor Speedway
- *c* David Taylor / Allsport
- *b* Pascal Rondeau / Allsport

## Introduction

Welcome to the exciting (and noisy) world of auto racing, a sport that is attracting the interest of youngsters in rapidly increasing numbers. Stock car racing, one of the many types of high-speed competitions featured in this book, is the fastest-growing spectator sport of the 1990s. Millions of fans drive to racetracks around the country, and millions more watch events on television. The profusion of magazines, books, trading cards, models, apparel, and souvenirs further signifies the sport's explosive growth and popularity.

An article on mathematics education reform in a recent issue of *Phi Delta Kappan* began, "Mathematical power for all students! Scientific and technological literacy, communication and higher problem-solving skills are the notes that make up that rallying cry." *RacingMath* was written with an eye toward linking these objectives to motivating, "real world" assignments. The book begins with specific, single-page activities, followed by more open-ended projects that typically involve research and cooperative learning. Throughout, the emphasis is on *how* and *why,* the essence of problem-solving. Many activities and projects require the understanding and use of several math skills, and many call on reading, writing, and thinking skills.

The book's grade level range is 4–8, but there are no levels assigned to specific activities or projects. Aptitudes and abilities vary within each grade as well across grade levels. The authors of this book feel strongly that the best person to determine an assignment's appropriateness is the teacher.

Auto racing is filled with involving, contextualized math "events." Youngsters love to compare the zero-to-sixty-miles-per-hour performances of their favorite cars, they enjoy calculating comparative lap times for stock car racers, and they are fascinated by speed records. Just imagine how drawn they'll be to an activity on "How to Be a Race Car Driver."

Depending on your students' proficiencies the assignments in this book can be used for enrichment, review, or reinforcement. The activities and projects progress from simple calculation exercises to more challenging involvements.

The authors make no apologies for the activities that invite readers to simply have fun. Many years ago John Dewey said, ". . . experience is most rewarding when it involves the seemingly contradictory traits of rigor and playfulness." Mathematics is a means of investigation, a way of solving problems, and a way of thinking. It is connected to everything. Auto racing provides a rich connection.

**Contents**

# Contents

| | | |
|---|---|---|
| 1 | **ACTIVITIES** | |
| 2 | **Road race** | subtraction |
| 3 | **Magazine madness** | addition |
| 4 | **Speeding up** | estimation |
| 5 | **Autocross ladies** | reading and writing decimals |
| 6 | **On the road** | reading charts and graphs |
| 7 | **Winston Cup race** | estimation |
| 8 | **Fast laps** | multiplication and division |
| 9 | **Favorite numbers** | charts and graphs |
| 11 | **Points to win** | charts and graphs |
| 12 | **Drag racing** | interpreting percents |
| 13 | **Day at the racetrack** | writing a story question |
| 14 | **One-handed driving** | division, averaging |
| 15 | **Odd ovals** | division, averaging |
| 16 | **Model money** | multiplication, division |
| 17 | **Indy 500 vs. U.S. 500** | charts and graphs |
| 18 | **Hare racing** | making projections and percents |
| 19 | **Pikes Peak Hill Climb** | decimal computation |
| 21 | **Way graphic** | averaging |
| 22 | **Racing suits and safety** | reading charts, division |
| 23 | **And laundry too** | reading charts, addition |
| 24 | **Gross tongues** | percents, computation |

| | | |
|---|---|---|
| 25 | **Long day** | addition, multiplication |
| 27 | **Speedway capacity** | multiplying whole numbers and decimals |
| 28 | **Service with a smile** | reading graphs, ratios |
| 29 | **Points, points, points** | reading and interpreting data |
| 31 | **Staggering around** | reading and interpreting data |
| 33 | **Michael Andretti vs. Al Unser, Jr.** | reading and interpreting data, ratios and percents |
| 35 | **Metric chassis** | metric conversions |
| 36 | **First wins** | computing average and median |
| 38 | **Have a seat** | measurement and computation |
| 39 | **French Grand Prix** | calculating percentages |
| 41 | **Road rally** | interpreting data, averaging |
| 42 | **Big winnings** | addition, calculating median |
| 44 | **Spider, spider** | multiplying decimals, percentages |
| 45 | **Television ratings** | interpreting data, calculating percentages |
| 46 | **Stock car growth** | calculating percentages |
| 48 | **Not a drag** | calculating percentages |
| 49 | **Victory's Vintage Storage** | interpreting data |
| 51 | **Tight turns** | geometry |
| 53 | **Home sweet track** | percents computation |
| 54 | **Michigan International Speedway** | interpreting data, averaging |

| | | |
|---|---|---|
| 55 | **Track size** | *geometry* |
| 56 | **Legends cars** | *calculating fractions* |

| | | |
|---|---|---|
| 57 | **PROJECTS** | |
| 58 | **So you want to be a race car driver** | *estimation, calculation* |
| 60 | **Fast, faster, fastest** | *communication, reasoning* |
| 61 | **Follow your favorites** | *calculation, prediction* |
| 64 | **The Brickyard** | *calculation, percentage* |
| 66 | **British Grand Prix** | *conversion* |
| 67 | **World land speed records** | *computation, reasoning* |
| 68 | **Who's Who directory** | *research, communication* |
| 71 | **Winston Cup series** | *interpreting data, communication* |
| 73 | **Shirley Muldowney** | *research, communication* |
| 74 | **Ice racing** | *calculation, reasoning* |
| 75 | **Indianapolis 500 statistics** | *making a graph, estimating, writing* |
| 80 | **Engine blocks and blockheads** | *writing, making a chart* |
| 81 | **Hot rod season** | *calculation, reading charts* |
| 85 | **Autocross racing** | *research, measurement* |
| 87 | **Hot rods and funny cars** | *calculation, research, writing* |
| 88 | **Le Mans 24-hour statistics** | *creating charts, percentages* |
| 92 | **Pistons and power** | *research, writing* |
| 93 | **Daytona 500 NASCAR statistics** | *calculating averages, percentages, drawing conclusions* |
| 96 | **Pikes Peak or bust** | *research, creating charts* |
| 98 | **Fabulous Ferraris** | *percentages, ratios* |
| 100 | **Answer key** | |

# ACTIVITIES

Activities

# Road race

A total of 60 drivers entered the Kiltarten Road Race. The cars traveled over a 3-mile road course for 24 laps.

1. By the end of the third lap, 4 cars blew their engines. How many cars were left in the race?

2. During the tenth lap, a 7-car pileup eliminated each car involved in the crash. How many cars were left?

3. In the fifteenth lap, corner workers waved 2 cars out of the race. How many cars were left?

4. An 11-car pileup in the twentieth lap eliminated each car involved in the crash. How many cars were left?

5. How many cars were eliminated in all?

# Magazine madness

Kevin and Kara love to read about all kinds of auto racing. They especially love to read about it in magazines.

| Popular Hot Rodding | $16.94 |
| --- | --- |
| Stock Car Racing | $28.00 |
| Autoweek | $32.00 |
| Circle Track & Racing Technology | $23.95 |
| Indy Car Racing | $29.95 |

**1.** How much will it cost Kevin to subscribe to both *Indy Car Racing* and *Popular Hot Rodding*?

**2.** How much will it cost Kara to subscribe to both *Circle Track & Racing Technology* and *Autoweek*?

**3.** What is the least Kara could spend by subscribing to 2 magazines?

**4.** What is the most Kevin could spend by subscribing to 2 magazines?

**5.** How much would it cost to subscribe to all 5 magazines for a year?

RacingMath

# Speeding up

The Indianapolis 500 is one of the most famous auto races in the world. The winner drives the 500-mile race in the shortest time. In 1915 Ralph DePalma won in 5 hours and 34 minutes (5:34). His average speed was 90 miles per hour. Twenty years later, in 1935, Kelly Petillo won in 4:42 at an average speed of 106 miles per hour.

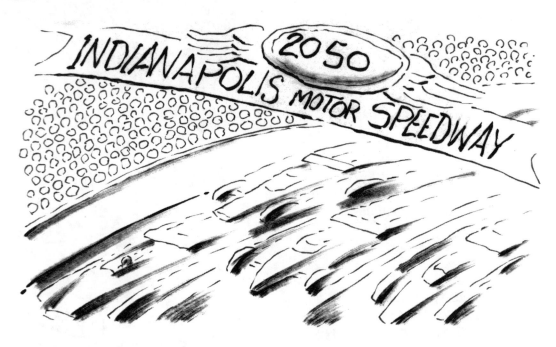

1. Bob Sweikert won the 1955 Indy 500. His winning time was probably

   3:54     5:10     1:20

2. Sweikert's winning speed was probably

   170 mph     110 mph     128 mph

3. The 1975 and 1976 Indy 500s weren't a full 500 miles because of rain. In 1977 A. J. Foyt won. His winning time was probably

   3:54     4:48     3:05

4. Foyt's winning speed was probably

   161 mph     255 mph     115 mph

5. In 1990 Arie Luyendyk won in 2:41. His winning speed was probably

   300 mph     186 mph     292 mph

RacingMath

# Autocross ladies

Both men and women race in autocross meets sponsored by the Sports Car Club of America. In an autocross, drivers race their cars through a course of orange cones. Each cone a driver hits adds a penalty of 2 seconds to his or her score. Samantha, Lauren, and Kelly raced in a ladies' class autocross.

**1.** Kelly's slowest run was 49.0 seconds. Her fastest was 44.2 seconds. What was the difference between the 2 runs?

**2.** Samantha's fastest run was 43.8 seconds, but she hit 2 orange cones. What was her score for that run in seconds?

**3.** Lauren's first run was a perfect but pokey 51.2 seconds. Her second run was 50.0 seconds and her third was 50.5 seconds. Lauren's fourth run, in which she hit 1 cone, was 48.9 seconds. What was her final score for the fourth run?

**4.** Which of the 3 women had the lowest score?

### Challenge problem

What was Lauren's average speed for her 4 runs? (Don't count the penalty.)

RacingMath

# On the road

Zack and Jessica's father is a stock car driver and their mother is a member of his pit crew. In the summer the family travels from race to race in a motor home. During one 2-week circuit, they went to 4 cities.

| Cities | Miles |
| --- | --- |
| Wichita to Springfield (MO) | 287 |
| Springfield to Memphis | 281 |
| Memphis to Corinth | 93 |
| Corinth to Chattanooga | 225 |
| Chattanooga to Atlanta | 117 |
| Atlanta to Charlotte (NC) | 240 |
| Charlotte to Raleigh | 153 |
| Raleigh to Savannah | 343 |
| Savannah to Jacksonville | 140 |
| Jacksonville to Baton Rouge | 603 |
| Baton Rouge to Oklahoma City | 611 |
| Oklahoma City to Wichita | 157 |

1. How far is it from Wichita to Memphis?

2. From Memphis to Atlanta?

3. From Atlanta to Raleigh?

4. From Raleigh to Jacksonville?

**Challenge problem**

After the race in Jacksonville, the family returned home to Wichita. How many miles in all did the family travel on this circuit?

**RacingMath**

# Winston Cup race

One of the races in the Winston Cup series is run at the Martinsville Speedway in Martinsville, Virginia.

1. The Martinsville Speedway is 0.526 mile around. The Winston Cup race covers 500 laps. The race is about

   500 miles    250 miles    350 miles

2. The winner completed the race in 3 hours, 13 minutes, and 50 seconds (3:13:50). He was traveling about

   120 mph    100 mph    80 mph

3. It took approximately 194 minutes to complete the race. This means that in 1 minute, the winner completed

   1 lap    more than 2 laps    more than 5 laps

4. The purse for the Winston Cup race was $1,219,566, with approximately 15% going to the drivers who placed first, second, and third. The remaining 85% was divided among the other 33 drivers. Together the top 3 drivers earned approximately

   $500,000    $175,000    $750,000

5. The first 3 finishers received approximately the same amount of money. Each driver received about

   $58,000    $250,000    $175,000

# Fast laps

The track illustrated below is 2.5 miles around.

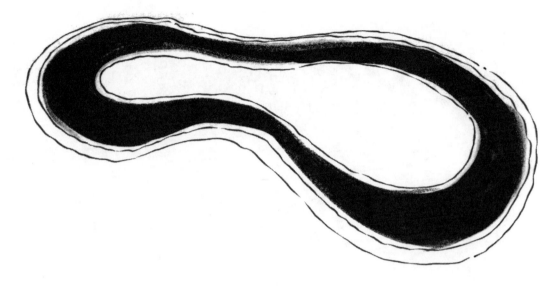

1. How far will a racing car go if it travels around the track 20 times?
   _____

2. How many laps will such a car have to travel to go 200 miles?
   _____

3. How fast will the car have to travel to go around the track in 1 minute?
   _____

4. If a racing car completed a 500-mile race in 2-1/2 hours, what would be its average speed?
   _____

5. Do you think such an average speed is possible? Explain why or why not.
   _____
   _____

# Favorite numbers

Latisha, Andre, Patrick, and Jasmine each have a favorite car number. The numbers are 7, 13, 22, and 80. Use the chart and the clues below to find each person's favorite number.

Clue 1  Latisha's number is a multiple of 10. Her number must be 80. Write YES next to 80 under Latisha's name. Then write NO in the other 3 boxes under her name. Write NO in the 80 row for Andre, Patrick, and Jasmine.

Clue 2  Both Andre's and Jasmine's favorite numbers are odd numbers. Only 7 and 13 are odd. So 22 must be Patrick's favorite number.

Clue 3  Jasmine's number is larger than Andre's. Finish the chart below so that you know everybody's favorite number.

|    | Latisha | Andre | Patrick | Jasmine |
|----|---------|-------|---------|---------|
| 7  |         |       |         |         |
| 13 |         |       |         |         |
| 22 |         |       |         |         |
| 80 |         |       |         |         |

RacingMath

Complete the chart below to find each person's favorite number.

Clue 1  Joel's favorite number will divide into any whole number and have no remainders.

Clue 2  Steffi's number is equal to 6 dozen.

Clue 3  Bart's number is 1/3 of Roxanne's number.

|    | Steffi | Joel | Roxanne | Bart |
|----|--------|------|---------|------|
| 1  |        |      |         |      |
| 18 |        |      |         |      |
| 54 |        |      |         |      |
| 72 |        |      |         |      |

# Points to win

The chart shows the standings of the 1996 PPG Indy Car World Series after 10 of 16 races were completed. Study the chart and answer the questions.

### PPG Indy Car World Series
*After 10 of 16 Rounds*

| Standings | Points |
|---|---|
| Jimmy Vasser | 102 |
| Al Unser, Jr. | 99 |
| Gil de Ferran | 92 |
| Christian Fittipaldi | 78 |
| Michael Andretti | 73 |
| Greg Moore | 60 |
| Scott Pruett | 60 |
| Alex Zanardi | 55 |
| Andre Ribeiro | 49 |
| Paul Tracy | 49 |

**1.** Suppose that the winner of the 1996 PPG Indy Car World Series finished with 160 points. How many points per race would he have averaged?

_____

**2.** What is Jimmy Vasser's average number of points per race after 10 races?

_____

**3.** Suppose that Michael Andretti won 4 of the remaining 6 races with an average of 20 points a win, and came in third in the last 2 races with an average of 14 points for third. And suppose that Vasser averaged 10 points per race for the remaining 6 races. What would Andretti's total be?

_____

What would Vasser's total be?

_____

**4.** If Gil de Ferran placed second in all 6 of the remaining races with an average of 17 points a race and if Vasser averaged 9 points per race for the remaining 6 races, who would win?

_____

**5.** Suppose that Al Unser, Jr., placed sixth, fifth, fourth, third, second, and first in the last 6 races with points of 8, 10, 13, 15, 18, and 21. What would his total points be?

_____

If Vasser averaged 8 points per race for the remaining 6 races, what would his total points be?

_____

**Racing Math**

# Drag racing

Travis and Ivan raced head-to-head 85 times last summer.

**1.** Travis won 65% of the races. How many did he win?

**2.** Travis's total winnings were $14,000, but $4,800 of that was in merchandise. What percent of Travis's earnings was in cash?

**3.** Ivan won $5,100, but 80% of that was in merchandise. How much did Ivan win in cash?

**4.** Ivan sold his new merchandise at a cut-rate price of 45% of its value. How much money did he make from the merchandise?

### Challenge problem

Travis and Ivan will go heads-up again next summer in another 85 drag races. If Ivan increases his number of wins by 40%, how many of the races will he win?

# Day at the racetrack

Here are the facts and here are the answers. Write the correct question to go with each answer.

**1.** Two adults and eight kids piled into a van and headed for the raceway. The adults paid $20 a ticket for grandstand seats and the kids paid $5 a ticket for grandstand seats. The answer is $80. What is the question?

_____

_____

**2.** The answer is $40. What is the question?

_____

**3.** There are 60,000 seats in the grandstands and another 15,000 on the hillside. The answer is 45,000. What is the question?

_____

**4.** The hillside seats cost $15 for adults and $4 for kids. The answer is $62. What is the question?

_____

_____

**5.** Every ticket to the race was sold. There were 12,000 kids in the grandstands and 4,500 on the hillside. The answer is 58,500. What is the question?

_____

_____

_____

**13** RacingMath

# One-handed driving

Tony Simon started racing when he was 8 years old. After racing professionally for many years, he retired from racing and became a truck driver.

**1.** When Tony Simon was 8 years old, he raced micro-midget cars on 1/8-mile ovals in southern California. In a 25-lap race, how many miles did Tony drive?

**2.** In 1971 Tony raced Sprint cars on dirt tracks and was voted Rookie of the Year after winning 6 main events. If each of his 6 wins occurred on a 5/8-mile oval and each consisted of 50 laps, how many miles did Tony drive in the 6 victories?

**3.** Racing at the Pennsylvania Fairgrounds in 1973, Tony lost control of his sprinter and it flipped. He was hospitalized with a fractured skull and broken collarbone. His wife Diane put Tony in the back of their motor home and drove home to California in 52 hours. If the distance from the hospital to their home was 2,777 miles, how many miles an hour did Diane average?

**4.** If she slept for 9 of the 52 hours, how many miles an hour did she average during her driving hours?

**5.** The following year, Tony broke both arms and smashed his right hand in another racing accident. Doctors had to amputate the hand. After that, Tony wore a hook in its place. In 1979 he won 5 major races and in 1982 he won 8. He retired in 1987 to drive trucks full time. If Tony won 39 major racing events in 17 years of racing, what was his average number of big wins per year?

# Odd ovals

Jermaine, Billy, Hector, and Courtney all raced on the same day on paved oval tracks. Each of them raced on a different track in a different race. The clues below will help you figure out who was where. Fill out the chart to show who raced on which track, at which distance, and at which speed.

| Driver | Track size | Race Distance | Average Speed |
|--------|------------|---------------|---------------|
|        | 0.62 miles |               |               |
|        | 1.20 miles |               |               |
|        | 1.50 miles |               |               |
|        | 2.50 miles |               |               |

1. Billy won his 200-mile race by completing all 80 laps in the fastest time.

2. Courtney placed tenth in a 300-mile race that she completed in 250 laps.

3. Hector completed 500 laps totaling 310 miles to take seventh place.

4. Jermaine placed sixth in a 375-mile race of 250 laps.

5. What was the average speed of each driver? Think about the size of the track. If all other conditions (such as weather) are equal, will a small track raise or lower the average speed? Pick each driver's average speed from the following numbers: 114mph, 106 mph, 142 mph, 91 mph.

### Challenge problem

Explain what the general connection is between the size of an oval track and the average speed of the cars racing on it. Draw diagrams of a small and large oval track to go along with your explanation.

# Model money

Twenty-First Lap produces model Indy Race Cars, each built to 1/21 the scale of the actual car. A 1:21 ratio means that each inch of the model car is equal to 21 inches of the actual Indy racer. After the 1997 Indy season, the company produced a limited edition of 5 models. It manufactured 2,100 cars of each model. Each car sold for $42.50.

**1.** Jeremy bought 1 of each of the 5 model cars. How much did this cost him?

**2.** How many cars did Twenty-First Lap produce in all?

**3.** If Twenty-First Lap sold all the 1997 models it produced, how much money did it receive?

**4.** If the production cost for each car was $13.25, how much profit did Twenty-First Lap make per car? Profit is the amount left after costs are subtracted.

**5.** Suppose that 21 years after Jeremy buys his model cars, they become collector's items, selling for $425 each. If Jeremy sold all 5 of his 1997 models, how much income would he receive? What would his profit be?

### Challenge problem

One of Jeremy's models is 8 inches long. How long is the real Indy racer that the model is based on?

RacingMath 16

# Indy 500 vs. U.S. 500

In 1996 many drivers pulled out of the Indianapolis 500 and raced instead at the U.S. 500 PPG Indy Car Championship in Michigan. How did the results of the 2 races match up?

|  | Lap length | Average speed | Margin of victory | Fastest lap | Fastest qualifier |
|---|---|---|---|---|---|
| Indianapolis | 2.5 mi | 147.956 mph | 0.695 sec | 38.119 sec | 236.986 mph |
| Michigan | 2.0 mi | 156.403 mph | 10.995 sec | 30.836 sec | 232.025 mph |

**1.** How much longer was each lap at Indianapolis?

**2.** How much faster was the average speed of the winner at Michigan?

**3.** How much wider was the margin of victory at Michigan?

**4.** How much faster was the fastest lap at Michigan?

**5.** How much faster was the fastest qualifier at Indianapolis?

### Challenge problem

Which race do you think was the most exciting? Why?

# Hare racing

Harry owns his own business, Hare Racing. He sells auto supplies. Harry buys his products for a wholesale price and marks them up a certain percentage.

**1.** If Harry buys his least-expensive helmets for $62 each and marks them up 300%, what does he sell them for?

_____

**2.** His most-expensive helmet costs him $165 and he marks it up 350%. What does he sell this helmet for?

_____

**3.** A deluxe tire gauge costs Harry $8.75. He marks it up 400%. What does it sell for?

_____

**4.** Hare Racing sells its least-expensive driving suit for $480. If Harry marked the suit up 200%, what did he pay for it?

_____

**5.** Hare Racing sells a custom-designed driving suit for $1,299. If Harry marked up the suit 250%, what did he pay for it?

_____

### Challenge problem

Carlene purchased one of Harry's least-expensive helmets, a deluxe tire gauge, and a custom-designed driving suit. What did Carlene pay for all these items together?

_____

**Racing Math**

# Pikes Peak Hill Climb

The Pikes Peak International Auto Hill Climb is the second-oldest auto race in the United States (Only the Indianapolis 500 has been around longer.) Cars, trucks, tractor-trailers, and motorcycles compete in different classes. All of them race up a 12.42-mile gravel road to the top of Pikes Peak. In solving the problems below, remember that there are 60 seconds in a minute.

**1.** In 1994 New Zealander Rod Millen set a Pikes Peak record by racing to the top in 10:04.06—10 minutes and 4.06 seconds. In 1995 Pikes Peak was so icy on race day (July 4) that the race had to be shortened. In 1996 Millen won again, this time reaching the top in 10:13.64. How many seconds longer did his 1996 climb take than his 1994 climb?

**2.** In his 1996 victory, Millen drove a turbocharged, winged, all-wheel-drive Toyota. In his qualifying run to Glen Cove, Millen had a time of 4:28.17. Glen Cove is halfway to the top. How many miles did Millen cover in his qualifying run?

RacingMath

**3.** If he ran the first half of the actual race in 1996 in exactly the same time as he ran the qualifying run, how long did it take Millen to race the last half of the climb?

**4.** Robby Unser won the 1996 pickup truck class by racing to the top in 11:56.80. How much slower than the winning car was the winning truck?

**5.** The winner of the semi (tractor-trailer) division was Sid Compton, who reached the top in 15:48.88. How much more time did the semi take than the pickup to reach the top?

# Way graphic

Every summer Gina travels with her father, who races stock cars. Gina makes money by applying vinyl graphics to stock cars. She charges $9.50 to apply a graphic on a clean car. If the car needs to be washed, Gina charges $22 for the wash and graphic.

**1.** If Gina works on 8 cars a day, 4 of which are clean and 4 of which are dirty, how much money does she average each day?

**2.** Gina works 3.5 days a week during the summer. Based on the number of cars in Question 1, how much money does she average for the week?

**3.** Last summer Gina worked 18 weeks, earning her average each week. How much money did she make in all?

**4.** Last winter, Gina created vinyl graphics for drivers who didn't want to make their own. She charged $48.75 to create a graphic. Gina worked for 12 winter weeks and created an average of 2.5 graphics per week. How much money did she average per week?

**5.** How much money did she make over the winter?

**6.** Gina worked a total of 30 weeks (18 summer, 12 winter). What was her weekly average income for those 30 weeks?

# Racing suits and safety

Race car drivers run the risk of single-car and multiple-car accidents. Some of these accidents result in fires. The best protection against fire started by a crash is wearing fire-resistant underwear, a fire-resistant suit, and fire-resistant gloves, socks, shoes, balaclava, and full-face helmet. The National Fire Fighters Association and the U.S. military rate fire-resistant cloth by giving it a TPP number. TPP stands for Thermal Protective Performance.

| Suit | Cost | TPP Rating | Seconds before skin blisters | Cost per second |
|---|---|---|---|---|
| Red | $415 | 14 | | |
| Blue | $529 | 19 | | |
| Green | $788 | 60 | | |
| Yellow | $1,342 | 120 | | |

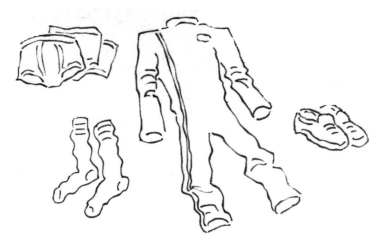

1. Once a fire-resistant suit ignites, a driver has only seconds before the flames make his skin blister. To figure how many seconds a driver has, divide the TPP rating by 2. Write the number of seconds for each of the 4 suits.

2. Fire-resistant racing suits are expensive. But they are a very important investment for each driver. The cost of the racing suits above is listed in dollars. But what is the cost in dollars per second? Divide the cost of the suit by the seconds a driver has before his skin starts to blister. Write the cost per second of each suit in the correct column.

RacingMath 22

# And laundry too

Formula One race crews dress in the colors and uniform of the company that sponsors the team. The cost of outfitting such a team is quite high. The £ sign below stands for pound, a British unit of money.

| Crew | Number of persons | Cost per person in British pounds | Cost per person in U.S. dollars |
|---|---|---|---|
| Race team | 28 | £1,439 | |
| Pit crew | 22 | £1,895 | |
| Drivers | 2 | £10,240 | |

**1.** How much did the sponsor pay in all for the race team uniforms (shorts, polo shirt, trainers, etc.)?

**2.** How much did the sponsor pay in all for the pit crew uniforms (fireproof helmets, gloves, boots, etc.)?

**3.** Each driver needed 3 made-to-measure race suits and 2 helmets. What was the sponsor's total cost for the 2 drivers?

**4.** What was the sponsor's total cost for all the uniforms?

**5.** After each race, the sponsor pays £1,155 to have all the uniforms laundered. If the sponsor enters 17 F1 races a year, what is its total laundry bill?

### Challenge problem

Assume that £1 = $1.60. Fill in the chart in U.S. dollars.

**RacingMath**

# Gross tongues

Some types of racing cars can be driven to a race, but most must be towed. To tow a car safely, a driver must understand the relationship between his gross trailer weight (GTW) and tongue weight (TW). The GTW is the combined weight of the trailer and its load. The TW is the downward force placed on the hitch ball by the trailer tongue. Each trailer manufacturer provides the TW—the driver must do the math! An ideal tongue weight is 10% of the gross trailer weight. The tongue weight should not exceed 15% of the gross trailer weight.

**1.** Darryl's car weighs 2,340 pounds and his trailer weighs 450 pounds. What is the ideal tongue weight for this combination?

**2.** What is the maximum tongue weight?

**3.** Michael's car weighs 1,715 pounds and his trailer weighs 278 pounds. What is the ideal tongue weight for this combination

**4.** What is the maximum tongue weight?

**5.** If Darryl and Michael buy a 2-car trailer that weighs 623 pounds and tow their cars on this trailer, what is the maximum tongue weight for the combination?

RacingMath 24

# Long day

Take a look at the typical day of a stock car driver and her pit crew:

| | |
|---|---|
| 5:30 A.M. | Pit crew of 2 loads car and leaves for speedway |
| 6:15 | Pit crew arrives at speedway |
| 8:00 | Becky arrives |
| 9:30 | Becky drives 10 test laps |
| 10:15 | Crew works on brakes |
| 11:15 | Becky drives 20 laps at full speed |
| 12:00 | Crew works on engine and chassis adjustments |
| 12:30 P.M. | Becky drives 10 test laps |
| 1:15 | Crew and Becky work on shocks and tires |
| 2:30 | Becky drives 7 more test laps |
| 3:30 | Crew works on engine and resets tire pressure |
| 4:15 | Becky drives 5 test laps |
| 5:00 | Pit crew and Becky lock up car and equipment for the day |

1. How many hours did the pit crew work?

   _____

2. How many hours did Becky work?

   _____

3. If each crew member is paid an hourly wage of $18.65 an hour, how much did each member of the pit crew make for the day?

   _____

4. If the crew members work this same schedule 3 days a week, how much does each crew member make in a week?

   _____

5. If Becky wins a $2,900 first-place prize in the next day's race and uses that money to pay the 2 members of her pit crew for their week's work, how much will she have left over for herself and other expenses?

   _____

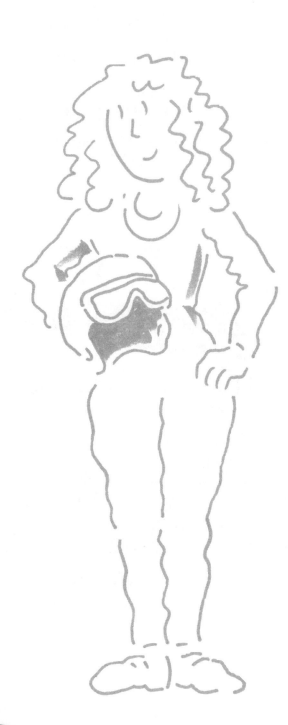

# Speedway capacity

Stock car racing is a very popular sport. Huge speedways have a seating capacity far greater than that of most football or baseball stadiums.

**1.** Trump Motor Speedway in Bridgeport, Connecticut, has a seating capacity of 80,000. If tickets for Trump cost $20 each, how much money does the speedway take in on a sellout event?

**2.** Homestead Motorsports Complex in Homestead, Florida, has a seating capacity of 65,000. If Homestead wants to gross the same amount as Trump collects for a sellout crowd, how much should Homestead charge for tickets? Round off this answer and others to the nearest 10 cents.

**3.** Texas Motor Speedway in Fort Worth has a seating capacity of 163,000. If it wants to collect the same gross amount as Trump, what will it charge for tickets?

**4.** If Texas Motor Speedway wants to make 50% more than the Trump gross amount, what will it charge for tickets?

### Challenge problem

California Speedway in Fontana has a seating capacity of 107,000. If it charges $22.50 per ticket, will it take in more or less money than the Texas Motor Speedway if the Texas Speedway wants to make 50% more than the Trump?

How much more money or how much less?

# Service with a smile

How fast does auto repair service take? It depends on who's doing it.

| Repair | Pit crew | Garage mechanic | Ratio |
|---|---|---|---|
| Replace windshield | 5 minutes | 3 hours | |
| Change 4 brake pads | 8 minutes | 2 hours | |
| Change springs and shocks | 15 minutes | 6 hours | |
| Change engine | 75 minutes | 14 hours | |

**1.** It takes a garage mechanic 180 minutes to replace a windshield; it takes the pit crew 5 minutes. The relationship of the mechanic's work time to that of the pit crew can be expressed as a ratio of 180:5. Express each of the relationships above as a ratio, putting the mechanic's time first.

**2.** How many times slower is the mechanic in doing each task?
   Replace windshield
   _____
   
   Change 4 brake pads
   _____
   
   Change springs and shocks
   _____
   
   Change engine
   _____

**3.** Assume that there are 4 members in the pit crew, but only 1 mechanic. For a true ratio, perhaps the mechanic's time should be divided by 4. Keeping the new ratio in mind, how many times slower is the mechanic in doing each of these tasks?
   Replace windshield
   _____
   
   Change 4 brake pads
   _____
   
   Change springs and shocks
   _____
   
   Change engine
   _____

### Challenge problem

On which of the tasks is the mechanic most efficient? Explain why you think this is so.

# Points, points, points

Study the chart below, answer the questions, and reach a conclusion.

### Indy Car World Series, 1994

| Driver | Points | Starts | Finishes | Laps led | Laps completed | Miles completed |
|---|---|---|---|---|---|---|
| Al Unser, Jr. | 225 | 16 | 12 | 677 | 1,954 | 3,358.862 |
| Emerson Fittipaldi | 178 | 16 | 11 | 404 | 1,966 | 3,368.019 |
| Paul Tracy | 152 | 16 | 9 | 503 | 1,673 | 2,855.538 |
| Michael Andretti | 118 | 16 | 9 | 166 | 1,746 | 2,922.498 |
| Robby Gordon | 104 | 16 | 12 | 59 | 1,780 | 3,047.470 |

**1.** Who completed the most miles?

**2.** Which driver completed the most laps?

**3.** Who had the most finishes?

**4.** Which driver led the most laps?

**5.** Who was second in the laps-led category?

**6.** Does the chart explain in any way how points are awarded in the Indy Car World Series?

**7.** Does the chart tell how many first-place, second-place, or third-place finishes a driver had?

RacingMath

**8.** Do you think points might be awarded on the basis of how a driver placed in a race? (First, second, third, fourth, fifth, and so on.)

_____

**9.** Explain what conclusion you draw about the fact that Al Unser, Jr., won more points than Emerson Fittipaldi did. What did Unser do that gave him more points?

_____
_____

# Staggering around

All 4 tires on an ordinary street automobile usually have the same amount of air pressure: for example, 30 pounds per tire. But for auto races around an oval track, it's essential that the air pressure in all 4 tires not be equal.

When cars race counterclockwise on an oval track, the driver is always steering to the left. To help the car stay on the track in a leftward direction, the driver wants stagger in his tires. Stagger is the difference in size between the 2 rear tires or the 2 front tires. The size is the circumference (or distance around) of each tire. Stagger comes from a difference in air pressure inside each tire. The less air pressure there is, the smaller the circumference of the tire.

| Driver | Right rear tire | Left rear tire | Stagger |
|---|---|---|---|
| Martinez | 86" | 85-1/4" | |
| Carter | 85-1/2" | 84" | |
| Dankovich | 86-1/4" | 85-3/4" | |
| Tweet | 86-1/2" | 84-3/4" | |
| Peebles | 85-1/4" | 83-3/4" | |

**1.** Fill in the stagger for each driver's rear tires.

**2.** Which driver has the least amount of stagger?

**3.** Which driver has the greatest amount of stagger?

RacingMath

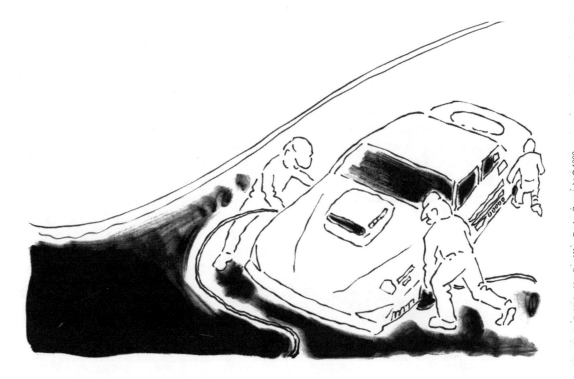

### Challenge problem

Try to picture the race car drivers going around an oval track counterclockwise. In each case, the "softer" tire is the left rear tire. For a very tight oval track, will a driver want more stagger or less stagger? Why?

_____

_____

# Michael Andretti vs. Al Unser, Jr.

Michael Andretti and Al Unser, Jr., race against other drivers on the Indy Car circuit. But they also compete against each other. To see how they match up, look at the statistics.

## Al Unser, Jr.
*Indy Car record*

| Year | Starts | Wins | Poles |
|------|--------|------|-------|
| 1982 | 1 | 0 | 0 |
| 1983 | 13 | 0 | 0 |
| 1984 | 16 | 1 | 0 |
| 1985 | 15 | 2 | 0 |
| 1986 | 17 | 1 | 0 |
| 1987 | 15 | 0 | 0 |
| 1988 | 15 | 4 | 0 |
| 1989 | 15 | 1 | 1 |
| 1990 | 16 | 6 | 1* |
| 1991 | 17 | 2 | 0 |
| 1992 | 16 | 1 | 1 |
| 1993 | 16 | 1 | 0 |
| 1994 | 16 | 8 | 4* |
| 1995 | 16 | 4 | 0 |
| Total | | | |

*Won PPG Indy Car World Series Championship

## Michael Andretti
*Indy Car record*

| Year | Starts | Wins | Poles |
|------|--------|------|-------|
| 1983 | 3 | 0 | 0 |
| 1984 | 16 | 0 | 0 |
| 1985 | 15 | 0 | 0 |
| 1986 | 17 | 3 | 3 |
| 1987 | 15 | 4 | 2 |
| 1988 | 15 | 0 | 1 |
| 1989 | 15 | 2 | 2 |
| 1990 | 16 | 5 | 4 |
| 1991 | 17 | 8 | 8* |
| 1992 | 16 | 5 | 7 |
| 1993 | Competed in Formula One | | |
| 1994 | 16 | 2 | 0 |
| 1995 | 17 | 1 | 3 |
| Total | | | |

*Won PPG Indy Car World Series Championship

**RacingMath**

**1.** Fill in the bottom of the chart by totaling each driver's starts, wins, and pole positions.

**2.** What is the ratio of pole positions to wins for

Andretti_____

Unser, Jr._____

**3.** What is the ratio of Indy Car wins to starts for

Andretti_____

Unser, Jr._____

**4.** Change the above ratios to percents, to show which percent of their Indy Car races each driver won:

Andretti_____

Unser, Jr._____

**5.** How many years did each driver compete in Indy Car Races from 1982 to 1995?

Andretti_____

Unser, Jr._____

**6.** What is the average number of races each driver won per year?

Andretti_____

Unser, Jr._____

### Challenge problem

Michael Andretti or Al Unser, Jr.—who do you think is the better Indy Car driver? Explain why.

_____
_____
_____
_____
_____
_____
_____
_____
_____
_____
_____

# Metric chassis

Because auto racing is a global sport, most manufacturers of chassis, engines, tires, and other auto components use both the English and metric systems of measurement. Suppose that the Indy Racing League, which releases chassis specifications for their cars, released these specifications only in metric. Make the conversions below.

**Conversion table**

1 millimeter = 0.04 inch

1 liter = 0.26 gallon

**1.** The length of the chassis must not be less than 4,750 millimeters *(mm)*. What is the minimum length in inches?

_____

**2.** The length must not be more than 4,875 mm. What is the maximum length in inches?

_____

**3.** The width must be 1,962.5 mm. This is how many inches?

_____

**4.** The wheel base must be 2,400 mm. How wide is this in inches?

_____

**5.** The gas tank holds 133 liters. How much is this in gallons?

_____

RacingMath

# First wins

How long does it take a Formula One driver to win his first Grand Prix? Sometimes 1 race is all it takes — sometimes 100 races aren't enough.

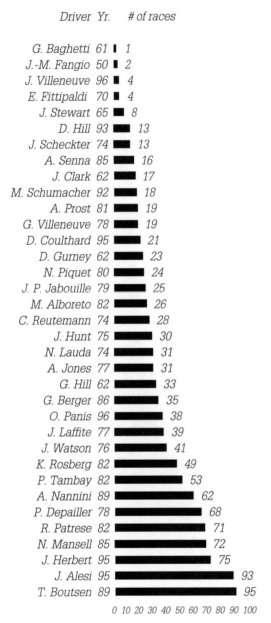

1. The chart shows how long it took some drivers to win their first Grand Prix. To the left of Baghetti's name, write the number 1. Continue numbering all the way through Boutsen. There are how many drivers represented? _____

2. In mathematics, the median is the middle number or figure in a series containing an odd number of items. For example, of the 5 numbers 4, 7, 8, 11, 12, the median is 8 — it is the middle number. In the list of drivers, which number is the median? _____

   There are how many drivers above the median? _____

   There are how many drivers below the median? _____

3. The median number of races run before a Grand Prix victory is what? _____

RacingMath 36

**4.** In mathematics, the average (also called the mean) is obtained by adding all the given numbers and dividing that sum by the total number of items. To find the average number of races run by the drivers before they finally won a Grand Prix, add all the races they drove before winning. The total is

_____

Divide that total by the number of drivers. The average number of races a driver ran before winning a Grand Prix is

_____

### Challenge problem

In real life, when might a median be a better number than the average (mean) to summarize a set of data? When might the average be a better number than the median to summarize a set of data?

_____

_____

_____

# Have a seat

Auto racers usually install special seats in their cars, both for safety and for driving efficiency.

| Model | Height (h) | Width (w) | Seat depth (d) | Weight |
|---|---|---|---|---|
| Street | 33.4" | 20.6" | 21.5" | 13 lbs. |
| Mixed | 32.1" | 22.3" | 19.9" | 12 lbs. |
| Competition | 33.8" | 21.7" | 28.8" | 24 lbs. |

1. If the 3 models were stacked on top of one another, what would be their combined height?

2. Which model occupies the smallest area (w x d)?

   How many square inches is its area?

3. Which model occupies the largest area (w x d)?

   How many square inches is its area?

4. Which model occupies the largest volume (h x w x d)?

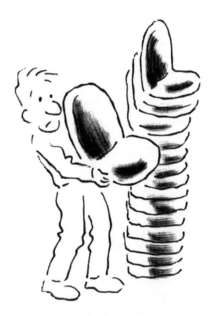

What is its volume in cubic inches?

### Challenge problem

What is the density (weight in ounces per cubic inch) of the heaviest model?

# French Grand Prix

Auto racing originated in France in 1894. The French Grand Prix is one of the world's most prestigious auto races.

| Pos. | Driver (Nationality) | Engine | Laps | Time/Status |
|---|---|---|---|---|
| 1 | Damon Hill (GB) | 3.0 Renault V-10 | 72 | 1:36.28.795 |
| 2 | Jacques Villeneuve (CAN) | 3.0 Renault V-10 | 72 | 1:36.36.922 |
| 3 | Jean Alesi (FRA) | 3.0 Renault V-10 | 72 | 1:37.15.237 |
| 4 | Gerhard Berger (AUS) | 3.0 Renault V-10 | 72 | 1:37.15.654 |
| 5 | Mike Hakkinen (FIN) | 3.0 Mercedes-Benz V-10 | 72 | 1:37.31.569 |
| 6 | David Coulthard (GB) | 3.0 Mercedes-Benz V-10 | 71 | Running |
| 7 | Olivier Panis (FRA) | 3.0 Mugen-Honda V-10 | 71 | Running |
| 8 | Martin Brundle (GB) | 3.0 Peugeot V-10 | 71 | Running |
| 9 | Rubens Barrichello (BRZ) | 3.0 Peugeot V-10 | 71 | Running |
| 10 | Mika Salo (FIN) | 3.0 Yamaha V-10 | 70 | Running |
| 11 | Ricardo Russet (BRZ) | 3.0 Hart V-8 | 69 | Running |
| 12 | Pedro Lamy (POR) | 3.0 Ford V-8 | 69 | Running |
| DNF | Heinz-Harald Frentzen (GER) | 3.0 Ford V-10 | 56 | Accident |
| DNF | Ukyo Katayama (JPN) | 3.0 Yamaha V-10 | 33 | Engine |
| DNF | Luca Badoer (ITA) | 3.0 Ford V-8 | 29 | Engine |
| DNF | Pedro Diniz (BRZ) | 3.0 Mugen-Honda V-10 | 28 | Engine |
| DNF | Jos Verstappen (NL) | 3.0 Hart V-8 | 10 | Accident |
| DNF | Eddie Irvine (GB) | 3.0 Ferrari V-10 | 5 | Gearbox |
| DNF | Giancarlo Fisichella (ITA) | 3.0 Ford V-8 | 2 | Fuel Pump |
| DNF | Andrea Montermini (ITA) | 3.0 Ford V-8 | 2 | Engine |
| DNS | Michael Schumacher (GER) | 3.0 Ferrari V-10 | 0 | Engine |
| DQ | Johnny Herbert (GB) | 3.0 Ford V-10 | 70 | No Time |

Pole: Schumacher, 1:15.989 (125.118 mph)
Winner's Averag Speed: 118.062 mph
Margin of Victory: 8.127 seconds
Fastest Lap: Villeneuve 1:18.610 (120.946 mph) on lap 48
Lap Leaders: Hill 1-27, 31-72; Villeneuve, 28-30
Note: All tires are Goodyear.
Legend: DNF=Did Not Finish, DQ=Disqualified, DNS=Did Not Start.

1. How many drivers were listed to start the race?

   _____

2. How many drivers finished the race?

   _____

3. What percent finished the race?

   _____

4. What percent were disqualified?

   _____

5. What percent did not start?

   _____

6. What percent failed to finish due to engine problems?

   _____

7. What percent were involved in accidents?

   _____

8. Of the finishers, what percent were driving cars with Renault engines?

   _____

9. What percent of the finishers were driving cars with Mercedes-Benz engines?

   _____

10. What percent of all engines were Ford engines?

    _____

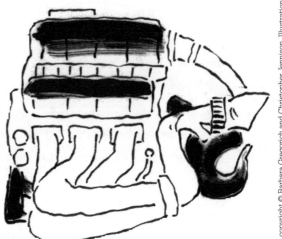

# Road rally

In a road rally, teams drive on public roads at legal speeds. Each team, consisting of a driver and a navigator, follows a set of printed instructions. The goal is to cover a certain distance at a certain speed — and arrive at the finish within a certain time.

|        | Distance (miles) | Speed | Hours | Minutes |
|--------|------------------|-------|-------|---------|
| Team A | 50               | 54    | .96   |         |
| Team B | 49               |       | .95   |         |
| Team C | 50               |       |       | 52      |

**1.** On the chart above, fill in the minutes and seconds for Teams A and B.

**2.** Fill in the hours for Team C.

**3.** Fill in the average speed for Teams B and C to 2 decimal points. (Remember that speed = distance ÷ time.)

**4.** If Team C covered the specified distance, which team failed to cover that distance?

_____

**5.** If the speed limit was 55 miles per hour, which team failed to observe the speed limit?

_____

RacingMath

# Big winnings

Below are the winnings of Indy Car's all-time top 20 drivers, based on records through 1994. Study the numbers, and then answer the questions that follow.

| Earnings | |
|---|---|
| Al Unser, Jr. | $15,379,906 |
| Emerson Fittipaldi | 13,272,875 |
| Bobby Rahal | 13,003,241 |
| Mario Andretti | 11,552,154 |
| Michael Andretti | 11,332,566 |
| Rick Mears | 11,050,807 |
| Danny Sullivan | 8,254,673 |
| Arie Luyendyk | 7,124,771 |
| Al Unser | 6,740,843 |
| A. J. Foyt | 5,357,589 |
| Raul Boesel | 5,273,584 |
| Scott Brayton | 4,500,711 |
| Tom Sneva | 4,392,993 |
| Roberto Guerrero | 4,275,163 |
| Scott Goodyear | 4,212,298 |
| Johnny Rutherford | 4,209,232 |
| Teo Fabi | 3,991,278 |
| Gordon Johncock | 3,431,414 |
| Paul Tracy | 3,389,817 |
| Nigel Mansell | 3,393,515 |

Indy Car's all-time top 20 drivers in victories, pole positions, and earnings, based on records through 1994.
From *The 1996 Information Please® Almanac.* Copyright © 1995 by Information Please LLC. All rights reserved. Reprinted by permission of INSO™ Corporation.

**1.** What is the difference in earnings between the first-place driver and the twentieth-place driver?

_____

**2.** What are the total winnings of the top 10 drivers?

_____

**3.** What are the total winnings of the drivers who placed eleventh through twentieth?

_____

**4.** What is the difference in earnings between the top 10 and the second 10?

_____

### Challenge problem

The median is the middle number in a series of numbers. The average (mean) of a series of numbers is obtained by adding all the given numbers and dividing that sum by the total number of items. What are the average (mean) earnings of a driver in the top twenty? What are the median earnings?

Average earnings: _____

Median earnings: _____

# Spider, spider

In 1995 Renault built both race-going versions and road-going versions of its Spider sports car. The racing Spider had a caged top with no windows or doors. Naturally, the road Spider had doors and windows. Discover some of the other differences between the 2 versions.

**1.** The engine of the road Spider has 150 horsepower at 6,000 revolutions per minute (rpm). The racer's engine has 1.2 times the horsepower. What is the racer's horsepower?

_____

**2.** The racer's engine has 1.083 times the road model's rpm. What is the racer's rpm?

_____

**3.** The road Spider weighs 2,050 pounds, but the racer weighs 91.4% of that. What is the racing Spider's weight?

_____

**4.** The top speed of the Spider racer is 113 miles per hour (mph). Imagine that the top speed of the road model is 85% of that. What is the road model's top speed?

_____

**5.** Imagine that the top speed of a powerful hot rod dragster is 300 mph. How many times faster is the hot rod than the Spider racer?

_____

# Television ratings

The A. C. Nielsen Company issues ratings for television shows. For major networks such as ABC, 1 rating point equals 1% of the potential viewing households (1.0 = 1%). One percent of the potential viewing households is 959,000 homes (1.0 = 959,000).

If a network show had a rating of 4.1, for example, then 4.1 x 959,000, or 3,931,900, equals the number of households viewing the program. The number of households is rounded to the nearest 100,000 and reported as a decimal. So 3,931,900 is reported as 3.9m. The m stands for million—3.9 million.

| Event | Network | Rating | Households |
|---|---|---|---|
| Indy Car/Miller 200 | ABC | 1.7 | |
| Indy Car/ITT Detroit Grand Prix | ABC | 2.2 | |
| NHRA Oldsmobile Springnationals | ABC | 1.6 | |
| IRL Indianapolis 500 | ABC | 6.6 | |
| NASCAR/Coca-Cola 600 | TBS | 4.6 | 3.1m |
| Indy Car/U.S. 500 | ESPN | 2.8 | 1.9m |
| NASCAR/UAW-GM Teamwork 500 | TNN | 3.9 | 2.5m |

**1.** Fill in the Households section of the first half of the chart above.

**2.** For cable networks such as TBS, ESPN, and TNN, the total number of potential viewing households varies. In each case, it is less than 959,000. To determine what 1% of each of these network's potential viewing households is, divide the rating into the number of households. What number of households equals 1% for each network?

TBS _____

ESPN _____

TNN _____

**RacingMath**

# Stock car growth

Stock car racing is becoming more and more popular every year. At one time, mostly men attended stock car races. Today, many of the fans are women. Kids attend, too, when families go to the races. The National Association for Stock Car Auto Racing (NASCAR) keeps track of the growth of the sport.

1. In 1975 girls and women made up 15% of the fans at stock car races. In 1995 they accounted for 38% of the fans. Of the 5,000,000 fans who attended NASCAR Winston Cup races in 1995, how many were females?

2. What was the average yearly percent increase in the number of female fans in the 20 years from 1975 to 1995?

**3.** If the same rate of increase continues, what percent of NASCAR fans will be females in 2005?

**4.** Remember that in 1995, five million fans attended NASCAR Winston Cup races. This was 300% of the number of fans who attended in 1980. How many attended in 1980?

**5.** If this rate of growth continues, how many fans will attend stock car races in 2010?

**6.** In 1995 the sale of NASCAR clothing, caps, and souvenirs totaled $600,000,000. If only the fans who attended races that year purchased these products, and if each fan bought something, what was each fan's average expenditure?

**7.** Is it likely that each fan spent this much money on merchandise?

**8.** What conclusions can you draw about who buys NASCAR merchandise and about the popularity of the sport?

**9.** If females purchased a percent of 1995 merchandise equal to their percent as race fans, how much of the $600,000,000 represents spending by males?

### Challenge problem

What was the average yearly percent increase in attendance for each of the 15 years from 1980 to 1995?

# Not a drag

In drag racing, two cars race against each other down a quarter-mile track. In order to achieve faster speeds, drivers modify their car engines.

1. Suppose that Lance's 112-horsepower (hp) Honda, with no modifications, can run the quarter mile in 16.5 seconds. Lance increases the hp to 150. This is an increase of what percent?

2. With the modified 150-hp engine, Lance's Honda can run the quarter mile in 15.1 seconds. This is a decrease in time of what percent?

3. If Lance's time decreased by the same percent that his horsepower increased, how fast would the 150-hp car do the quarter mile?

4. Too bad—the relationship between increased horsepower and decreased time is not that simple. Lance keeps making changes to his Honda. Along with other changes, he adds a turbocharger system that increases the hp from the original 112 to 230. This is an increase of what percent?

5. After all these modifications, Lance's Honda can run the quarter mile in 12.45 seconds. This is a decrease in time of what percent?

# Victory's Vintage Storage

Victory owns a unit of 15 garages connected to a Motorsports Complex. She rents the garages to owners of vintage race cars. Many of these cars cost over $500,000. Each renter has access to the racetrack. Each garage is heated, and the garage units have 24-hour access and 24-hour security. The complex itself has specialists who restore vintage cars.

## STARTER
Width 50'
Depth _____
Area 1,000 sq ft
Yearly _____
Monthly _____

## PIT STOP
Width _____
Depth 22'
Area 1,320 sq ft
Yearly _____
Monthly _____

## FINAL LAP
Width 65'
Depth _____
Area 1,625 sq ft
Yearly _____
Monthly _____

## CHECKERED FLAG
Width 70'
Depth 32'
Area _____
Yearly _____
Monthly _____

Victory's garages come in 4 sizes, which she has named Starter, Pit Stop, Final Lap, and Checkered Flag. Study the dimensions and area of each of the 4 garages. If the width or depth are given, fill in the area. If the width or depth are missing, fill them in.

Car owners who rent at Victory's Vintage Storage must sign a yearly contract. Victory charges $28 per square foot per year. Fill in the yearly rent on each garage.

Owners can pay monthly instead of yearly, but if they pay monthly, they must add a $55 processing fee to the monthly rent. Fill in the monthly rent for each garage.

# Tight turns

On many racetracks, turns pose the most difficult challenge. A 90° turn, which is usually made on city streets at a very slow speed, has to be executed at high speeds on a racetrack. Study the 2 racetracks and the information on angles, and then answer the questions that follow.

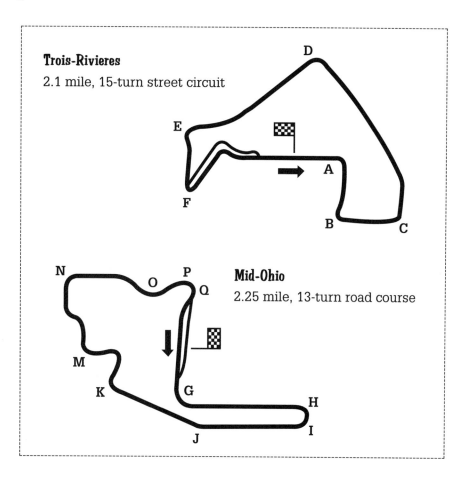

**1.** The turns labeled A, B, and C at Trois-Rivieres are all approximately what kind of angle?

**2.** D appears to be what kind of angle?

**3.** The turn labeled F is very sharp. It appears to be what kind of angle?

**4.** The turn labeled G at Mid-Ohio seems to be what kind of angle?

RacingMath

**5.** A driver completing a turn through points H and I has actually completed what kind of turn? (This type of turn is called a hairpin, from the way it looks.)

_____

**6.** Coming off this hairpin turn, the driver goes through J, which appears to be what kind of angle?

_____

**7.** Turns K, M, and N are all what kind of angle?

_____

# Home sweet track

Believe it or not, you can live in a condominium located inside the Charlotte Motor Speedway. The condos are actually expanded versions of the luxury skyboxes built into many football and baseball stadiums. Huge living room picture windows overlook the track. Down below, 700-horsepower engines propel stock cars at nearly 200 mph, producing a loud howling sound that sometimes lasts until midnight.

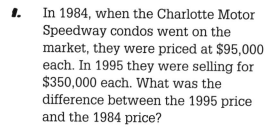

**1.** In 1984, when the Charlotte Motor Speedway condos went on the market, they were priced at $95,000 each. In 1995 they were selling for $350,000 each. What was the difference between the 1995 price and the 1984 price?

_____

Express the increase in price as a percent. To do that, divide the increase by the original price and express as a percent to the nearest whole number.

_____

**2.** What was the average annual increase over the 11-year period expressed as a percent?

_____

**3.** If this rate of increase continues for 10 more years, what will the condo be worth at the end of the 10-year period? To get the answer, multiply the 1995 price by the annual percent increase by 10 years. Then, don't forget to add this increase to the 1995 price.

_____

**4.** What will the condo be worth in 10 years if the rate of increase is only half as much?

_____

### Challenge problem

Imagine that you bought one of the trackside condos for $100,000. To do so, you took out a 20-year mortgage of $80,000 at an 8% per year interest rate compounded annually. (For every $100 you borrow at 8% compounded annually, you end up repaying $366.10.) Over 20 years, you would pay a total of _____ for the trackside condo (not counting the original $20,000 you paid before you took out the mortgage).

_____

If you decided to rent out your condo, how much rent would you have to charge per month to cover your mortgage and interest payment and also make a 10% profit every month?

_____

**53** RacingMath

# Michigan International Speedway

One lap, fifty laps, five hundred laps — what is the record at Michigan International Speedway, where the oval track is exactly 2 miles long? Round each final answer to 2 decimal points.

| Distance | Date | Driver | Speed (mph) | Time |
|---|---|---|---|---|
| 1 lap | 7-31-93 | Mario Andretti | 234.275 | |
| 250 miles | 9-28-86 | Bobby Rahal | 181.701 | |
| 500 miles | 8-5-90 | Al Unser, Jr. | 189.727 | |

**1.** How long did it take Andretti to travel 2 miles on the track? Fill in the chart above. To measure in seconds, remember that there are 3,600 seconds in 1 hour. Round the answer off to 2 decimal places.

**2.** Fill in the time for Rahal and Unser, Jr. Remember to express your time in hours, minutes, and seconds, as in 1:40:35.

**3.** If Andretti had driven the entire race at the speed of his record lap, what would have been his time for the 500-mile race?

**4.** If Unser, Jr., ran a lap at the same speed he ran the 500-mile race on August 5, 1990, how long would it take him to complete 1 lap?

**5.** In 1990, when Unser, Jr., won the 500-mile race, the pole position was won by Emerson Fittipaldi, whose record for 1 lap was 222.593 mph. How long did it take Fittipaldi to complete one lap?

RacingMath 54

# Track size

Many stock car races are run on oval tracks. Look at the oval track represented on this page. Determine its exact size by studying the clues.

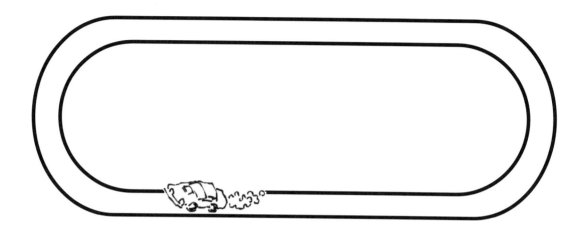

1. Each end of the track is a perfect semicircle. The size of each semicircle is the same. The radius of each semicircle is 1/9 of a mile. What is the length of 1 semicircle?

   _____

2. The 2 semicircles together equal what distance?

   _____

3. The 2 straightaways are equal to each other. The length of each straightaway is 4.5 times the radius of a semicircle. What is the length of 1 straightaway?

   _____

4. The 2 straightaways together equal what distance?

   _____

5. Therefore, what is the size of the track?

   _____

**RacingMath**

# Legends cars

Legends cars are designed as low-priced cars to be run economically on the many small tracks around the United States. The Legends cars are 5/8 the size of the old stock cars of the 1950s. Using Legends car specifications, you can determine the dimensions of the original stock cars.

### Legends car specs

| Width | 60" |
|---|---|
| Length | 126" |
| Height | 46" |
| Maximum weight | 1,175 lbs. |
| Maximum horsepower | 125 hp |
| Wheel diameter | 13" |

**1.** What was the width of the old stock cars of the 1950s?

**2.** What was their length?

**3.** Height?

**4.** Maximum weight?

**5.** Maximum horsepower?

**6.** Wheel diameter?

### Challenge problem

Legends cars were designed to race on tracks 4/10 mile or smaller. If 4/10 of a mile is 5/8 of the size of an average stock car track, what is the size of the average stock car track?

# PROJECTS

# So you want to be a race car driver

Maybe you dream of some day becoming another Terry Labonte, Al Unser, Jr., or Michael Schumacher. Many championship racing drivers began their careers at racing schools such as the ones whose locations are identified on the map below.

1. Which school is closest to your home?

2. Approximately how many miles is the school from your home?

3. How long will it take to drive from your home to the school if you average 50 mph?

4. If the car you travel in gets 25 miles to the gallon, approximately how much gas will you use on the trip?

5. With 2 or 3 classmates, prepare a report on one of the racing schools. Write to the school (addresses are on page 59) and ask for a catalog or information brochure. Your report can be written or presented to the class orally. Among the points you'll want to cover are minimum ages for students, types of racing instruction

given, types of cars used, instructors' qualifications, famous graduates, duration of the classes, the cost, and what the school will prepare you to do. Add illustrations of the cars used in the school and the track used by the school.

**6.** Imagine that you attended a racing school. Write a report describing the school, the cars, the instructors, the track, the other students, what you did at the school, and how it felt to be at the wheel of a racing car.

Addresses and phone numbers of racing schools:

Buck Baker Driving School
1613 Runnymede Lane
Charlotte, NC 28211
800-529-BUCK

Skip Barber Racing School
29 Brook St.
P.O. Box 1629
Lakeville, CT 06039
800-221-1131

Bob Bondurant School of High Performance Driving
Firebird International Raceway
20000 S. Maricopa Rd., Gate 3
Chandler, AZ 85226
800-842-RACE

Racing Adventures
P.O. Box 3960
Seminole, FL 34645
800-961-7223

Elf Winfield School
Franam Racing, Inc.
1409 S. Wilshire Drive
Minnetonka, MN 55305
612-541-9461

The Mid-Ohio School
Truesports, Inc.
94 North High St., Suite 50
Dublin, OH 43017
614-793-4615

**RacingMath**

# Fast, faster, fastest

If it has a motor, somebody will race it. People race cars, motorcycles, boats, and planes.

**World speed records**

| Vehicle | MPH | Driver | Date | Record set |
|---|---|---|---|---|
| Automobile | | | | |
| Motorcycle | | | | |
| Boat | | | | |
| Plane | | | | |

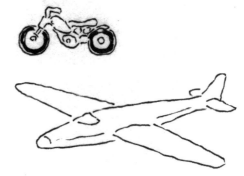

**1.** In your local or school library, use reference sources to find world speed records for automobiles, motorcycles, boats, and planes. Fill in the chart with these records.

**2.** Which form of motorized transportation is the slowest? Write 2 paragraphs explaining why this method is the slowest.

**3.** Choose 1 of the 4 vehicles and research the special modifications to the vehicle that enabled it to set a world speed record. Make a list detailing the special modifications: what each one is and how it enables the vehicle to go faster.

**4.** Create a board game in which the 4 vehicles are markers and the goal is to reach the finish line first by correctly answering math problems written on cards. Be as creative as you can with the markers, the board itself, the math problems, and the method of play.

RacingMath

# Follow your favorites

Check out your local newspaper for information on a racing series, such as the PPG Indy Car World Series. If you can't find this information in the newspaper, go to a library or newsstand to look at a weekly auto-racing magazine. Or, see if you can find point-standings information for your favorite racing series on the World Wide Web.

Read more about the series you have chosen to follow. Then fill out the information on the chart at the bottom of this page.

Name of Series _____

Dates of Series _____

Number of Races _____

Name of Trophy _____

Cash Prize to Winner _____

RacingMath

Choose 2 race car drivers competing in the series. Enter their names on the chart that follows. Each time one of the drivers competes in a race, enter his or her points for that race as well as cumulative points. Also enter where he or she finished in the race, as indicated on the chart.

| Name of race | Driver's name | | | Driver's name | | |
| --- | --- | --- | --- | --- | --- | --- |
| | Position finished | Points | Cumulative points | Position finished | Points | Cumulative points |
| | | | | | | |
| | | | | | | |
| | | | | | | |
| | | | | | | |
| | | | | | | |
| | | | | | | |
| | | | | | | |
| | | | | | | |
| | | | | | | |
| | | | | | | |
| | | | | | | |
| | | | | | | |
| | | | | | | |
| | | | | | | |

After the series you chose is finished, record the names, points, and prize money for the drivers who finished first, second, and third.

|  | Winners | Total points | Total prize money |
|---|---|---|---|
| 1st place |  |  |  |
| 2nd place |  |  |  |
| 3rd place |  |  |  |

**RacingMath**

# The Brickyard

In your school or town library, conduct some research on the Indianapolis 500. You can do this alone or with the help of 2 or 3 classmates. Answer each of the questions below.

1. Why is the race called the Indianapolis 500?

2. Why is the track nicknamed "The Brickyard"?

3. Are bricks still left there?

4. How long is the track, and how many laps are necessary to complete the race?

5. Who won the first Indianapolis 500, and what was his average speed?

6. Who won the last Indianapolis 500, and what was his average speed? State as a percent how much faster the last winner traveled than the first winner traveled.

7. How long did it take the first winner to complete the race?

**8.** How long did it take the most recent winner to complete the race? State as a percent how much faster he completed the race than the first winner did.

**9.** How many cars started the most recent race, and how many finished? State as a percent the ratio of cars finishing to those starting the race.

**10.** Who won the pole position for the most recent Indianapolis 500, and what was his qualifying time? Was this qualifying time faster or slower than the final winning speed? Why do you think that was so?

### Challenge problem

During the late 1960s, racing at Indianapolis changed significantly. Write a short report describing the change, how it came about, why, and what the results were.

**Racing Math**

# British Grand Prix

Throughout Europe and in much of South America and Asia, Grand Prix racing is highly popular. Study the British Grand Prix ticket prices below. What are these prices in the currency of other countries? To convert the currencies, you'll have to research the exchange rates. You might be able to find this information on the Internet, in a newspaper, or by calling a bank.

**Ticket prices**

|  | British pounds | German marks | Italian lira | Japanese yen | Canadian dollars | U.S. dollars |
|---|---|---|---|---|---|---|
| General | £30 | | | | | |
| Centre | £16 | | | | | |
| Grandstand north | £50 | | | | | |
| Grandstand south | £40 | | | | | |
| Children | £6 | | | | | |

# World land speed records

For decades, drivers set the world land speed record at Bonneville Salt Flats in Utah. Later, speed records were set at Black Rock Desert in Nevada. Learn more about land speed records at the library. Answer each of the questions below.

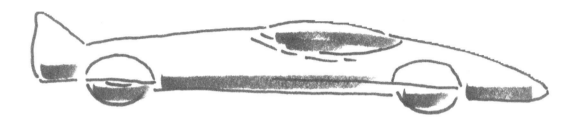

**1.** What was the earliest land speed record set at Bonneville? What was the most recent land speed record set there? Create a chart showing various Bonneville speed records.

**2.** What is the current land speed record and who holds it? Make a chart showing various land speed record holders: list their names, the names of their vehicles if possible, the place the record was set, the date of the record, and the record itself.

**3.** What is the speed of sound? Do you think that land speed records will ever break the sound barrier? Explain why.

**4.** How much distance is needed today for a specially built car to set a land speed record? Why?

**5.** Find pictures of several of the cars that have set land speed records. Describe special outward details about the cars that enable them to travel at such great speeds. On paper, draw a vehicle designed to break land speed records. Give it a name.

### Challenge problem

Write a story in which knowledge of world land speed records is crucial to the plot.

# Who's Who directory

Who's who in auto racing? Who's who in Formula One? or NASCAR racing? or drag racing? Learn who's who by creating your own Who's Who directory.

First, decide how many racing car drivers your directory will cover. Go for a minimum of 10, a maximum of 20.

Next, decide whether you want your directory to include the top drivers from several different kinds of auto racing or the top drivers in only 1 kind of auto racing. Once you've decided, you can title your directory something like *Who's Who in Formula One Racing,* or *Who's Who in Drag Racing.*

Read about the best drivers in the area you've chosen. Collect their pictures from newspapers and magazines. Get their vital statistics: height, weight, date born, and birthplace. Research their race statistics, too: championships, major races won, standings, and so on.

Once you've collected all the materials, think about the design of your directory. A well-designed directory is not like a scrapbook. In a scrapbook, information is pasted down every which way. In a directory, all the pages have the same type of information in the same place. Experiment with page designs by drawing blocks on a sheet of paper. Label the location of each photo: Driver, Car, Race Photo, and so on. Where the words will fall, draw squiggles across the blocks to indicate type.

In designing your directory, be as creative as you want. Include an outline of the racer's home state or home country, with a dot representing his hometown if you wish. Or reserve a space for the driver's sponsors, or the driver's pets—it's up to you!

After you've decided on a design, put your directory together. You may use 8 1/2" by 11" sheets, which are easy to find, or you may use larger photo-collection paper. Just be certain you'll have enough for the body and a cover.

Write or type the text on paper before you paste it on the pages. Make sure it fits the space you allowed for text. It's okay to have less text than you allowed for, but more text won't fit.

After your directory is finished and you've designed the cover, you can have copies of the directory made for your friends. Take your sheets to a copy shop, or ask someone to make photocopies for you.

# Winston Cup series

The charts that follow give you information on the 1995 Winston Cup NASCAR races. Study the charts and complete the projects below.

1. Create a chart showing the winners of the various 1995 Winston Cup races and their total earnings. Each winner's name should appear only once, with all the races he has won and his total earnings listed.

2. If the Winston Cup is awarded on the basis of who won the most money, then who won the 1995 Winston Cup? If it's awarded on the basis of who won the most races, who won the 1995 Winston Cup?

3. In a library, read about the 1995 Winston Cup winner. Write a brief report on the winner.

4. There are 28 races listed. What was the average amount of money won per race?

5. What was the average winning speed per race?

6. What do the identifying "400," "500," and "600" mean in the names of the Winston Cup races? Do they mean miles? laps? kilometers? Find out what the number means for each race by doing some library or Internet research. Then write a brief description of each race, giving its name, explaining what the identifying number means, and giving the size and shape of the track.

7. Choose any 2 of the race locations, as long as they are in different states, and read about the history of the host towns. Write several paragraphs describing each town, with an emphasis on its history. How much money does the race bring into the town?

# 1995 Winston Cup Series

| DATE | RACE | WINNER (position) | AVERAGE MPH | EARNINGS | POLE | QUALIFYING MPH |
|---|---|---|---|---|---|---|
| Feb. 19 | Daytona 500 | Sterling Marlin (3) | 141.710 | $300,460 | D. Jarrett | 193.498 |
| Feb. 26 | Goodwrench 500 | Jeff Gordon (1) | 125.305 | 167,600 | J. Gordon | 157.620 |
| Mar. 5 | Pontiac 400 | Terry Labonte (24) | 106.425 | 82,950 | J. Gordon | 124.757 |
| Mar. 12 | Purolator 500 | Jeff Gordon (3) | 150.115 | 104,950 | D. Earnhardt | 185.077 |
| Mar. 26 | TranSouth 400 | Sterling Marlin (5) | 111.392 | 86,185 | J. Gordon | 170.833 |
| Apr. 2 | Food City 500 | Jeff Gordon (2) | 92.011 | 67,645 | M. Martin | 124.605 |
| Apr. 9 | First Union 400 | Dale Earnhardt (5) | 102.424 | 77,400 | J. Gordon | 118.765 |
| Apr. 23 | Hanes 500 | Rusty Wallace (15) | 72.145 | 61,945 | B. Labonte | 93.308 |
| Apr. 30 | Winston 500 | Mark Martin (3) | 178.902 | 98,565 | T. Labonte | 196.532 |
| May 7 | Save Mart 300 | Dale Earnhardt (4) | 70.681 | 74,860 | R. Rudd | 92.132 |
| May 20 | The Winston Select | Jeff Gordon (7) | 148.410 | 300,000 | B. Labonte | 139.817 |
| May 28 | Coca-Cola 600 | Bobby Labonte (2) | 151.952 | 163,850 | J. Gordon | 183.861 |
| June 4 | Miller Draft 500 | Kyle Petty (37) | 119.880 | 77,655 | J. Gordon | 153.669 |
| June 11 | UAW-GM 500 | Terry Labonte (27) | 137.720 | 71,175 | K. Schrader | 163.375 |
| June 18 | Miller Draft 400 | Bobby Labonte (19) | 134.141 | 84,080 | J. Gordon | 186.611 |
| July 1 | Pepsi 400 | Jeff Gordon (3) | 166.976 | 96,580 | D. Earnhardt | 191.355 |
| July 9 | Slick 50 300 | Jeff Gordon (21) | 107.029 | 160,300 | M. Martin | 128.815 |
| July 16 | Miller Draft 500 | Dale Jarrett (15) | 134.038 | 72,970 | B. Elliott | 162.496 |
| July 23 | DieHard 500 | Sterling Marlin (1) | 173.187 | 219,425 | S. Marlin | 194.212 |
| Aug. 5 | Brickyard 400 | Dale Earnhardt (13) | 155.218 | 565,600 | J. Gordon | 172.536 |
| Aug. 13 | Bud at the Glen | Mark Martin (1) | 103.030 | 95,290 | M. Martin | 120.411 |
| Aug. 20 | GM Goodwrench 400 | Bobby Labonte (1) | 157.739 | 97,445 | B. Labonte | 184.403 |
| Aug. 26 | Goody's 500 | Terry Labonte (2) | 81.979 | 66,940 | M. Martin | 125.093 |
| Sept. 3 | Southern 500 | Jeff Gordon (5) | 121.231 | 70,630 | J. Andretti | 167.379 |
| Sept. 9 | Miller Draft 400 | Rusty Wallace (7) | 104.459 | 64,515 | D. Earnhardt | 122.543 |
| Sept. 17 | Delaware 500 | Jeff Gordon (2) | 124.740 | 74,655 | R. Mast | 153.446 |
| Sept. 24 | Goody's 500 | Dale Earnhardt (2) | 73.946 | 78,150 | J. Gordon | |
| Oct. 1 | Tyson Holly Farms 400 | Mark Martin (2) | 102.998 | 71,590 | T. Musgrave | 118.396 |

From *The 1996 Information Please® Sports Almanac.* Copyright © 1995 by Information Please LLC. All rights reserved. Reprinted by permission of INSO™ Corporation.

# Shirley Muldowney

Who is Shirley Muldowney? Find out through research. Then complete the projects below.

**1.** Write a "numbers report" on Shirley Muldowney. Describe her using as few words and as many numbers as you can. Write your report in such a way that a person who knows nothing about Muldowney will know a lot about her after reading your report.

**2.** Don Garlits raced at the same time as Muldowney. Research his statistics and create a chart comparing Muldowney and Garlits. Use mainly numbers to compare them.

**3.** Write a poem about Shirley Muldowney using ordinal and cardinal numbers as words in the poem. (If you don't know what ordinal and cardinal numbers are, look up the terms in a dictionary.)

**4.** The movie *Heart Like a Wheel* is based on Shirley Muldowney's life. Think of 5 other possible titles for such a movie, each including a cardinal or ordinal number. Write out the 5 titles.

# Ice racing

Did you know that every winter brings ice racing to Canada and parts of the United States? Automobiles race on the ice of frozen lakes and rivers, slipping and sliding all the way. Learn more about ice racing by researching the subject and answering the questions below.

1. The least expensive and most common form of ice racing is the rubber-to-ice category. Explain what rubber-to-ice means.

2. What is the metal-to-ice category?

3. If you wanted to drive as fast as possible on ice, which classification should you race in? Why?

4. One of the categories calls for greater driving skills and greater safety measures (roll bar and a 4-point safety harness) than the other. Which category requires greater driving skills and safety measures? Why?

5. To prepare a car for ice racing, a driver gets rid of all unnecessary weight in the vehicle. List 10 things you could remove from an old car to make it weigh less.

6. But, to drive well and win, a driver wants to add extra weight over the driving wheels (either front wheels or rear wheels). Explain the purpose of this extra weight. Define the word *traction*.

7. What is the weight of a cubic foot of water? Does ice weigh more or less than water? What is the weight of a cubic foot of ice?

### Challenge problem

In Canada and the northern part of the United States, it's not uncommon for cars or buses to travel on the ice of ponds or lakes — even on the Great Lakes. How thick does the ice have to be to support a 1-ton pickup truck?

# Indianapolis 500 statistics

The Indianapolis 500 has a long history. Use the chart on page 77 to learn more about it.

1. First, bring the Indianapolis 500 statistics up to date by filling in the information for each year from 1997 to the present.

2. Then determine the average winning speed in the Indianapolis 500 decade by decade. Use 1911 to 1919 as a decade labeled 1910s, use 1920 to 1929 as a decade labeled 1920s, and so on. Make a chart showing the average winning speed. Leave space on the chart for the projects below.

3. What was the average pole position speed, decade by decade? Put it on the chart you make.

4. Study the decade-by-decade statistics. Make a prediction about the average winning speed in the Indianapolis 500 for the next decade.

5. What, if any, is the relationship between the fastest qualifying speed (pole position speed) and the winning speed? Try to come up with a mathematical formula that expresses this relationship.

**6.** Create a collage of multiple Indianapolis 500 winners. Count up the multiple wins, find photos of the winners, and make the collage. Be certain there are numbers somewhere on the collage—either the number of multiple wins, or the average speeds, or something else.

**7.** Study the Indianapolis 500 statistics again. In which percent of cases does the pole sitter win the race?

**8.** Study the starting position of the winners. Make a grid that shows how many times each starting position has fielded a winner. Are some starting positions better than others? What are they?

**9.** Read about the Indy 500 starting positions. Imagine that you must write an article for your school newspaper explaining the starting positions. How are they organized? How are they determined? What is the configuration of these positions on the racetrack? Write an article that explains these things.

**10.** Watch the Indianapolis 500 on television, in person, or on a video. Write a poem about it.

### Challenge problem

What in the world is a Bowes Seal Fast Special? or a Ring Free Special? Read about an Indy special car from any year before 1950. Draw, trace, or photocopy a picture of the car. Create a chart that gives all the important information about the car.

# Indianapolis 500 statistics

| YEAR | WINNER (position) | CAR | MPH | POLE SITTER | MPH |
|---|---|---|---|---|---|
| 1911 | Ray Harroun (28) | Marmon Wasp | 74.602 | Lewis Strang | |
| 1912 | Joe Dawson (7) | National | 78.719 | Gil Anderson | |
| 1913 | Jules Goux (7) | Peugeot | 75.933 | Caleb Bragg | |
| 1914 | Rene Thomas (15) | Delage | 82.474 | Jean Chassagne | |
| 1915 | Ralph DePalma (2) | Mercedes | 89.840 | Howard Wilcox | 98.90 |
| 1916 | Dario Resta (4) | Peugeot | 84.001 | John Aitken | 96.69 |
| 1917–8 | Not held World War I | | | | |
| 1919 | Howdy Wilcox (2) | Peugeot | 88.050 | Rene Thomas | 104.78 |
| 1920 | Gaston Chevrolet (6) | Monroe | 88.618 | Ralph DePalma | 99.15 |
| 1921 | Tommy Milton (20) | Frontenac | 89.621 | Ralph DePalma | 100.75 |
| 1922 | Jimmy Murphy (1) | Murphy Special | 94.484 | Jimmy Murphy | 100.50 |
| 1923 | Tommy Milton (1) | H.C.S. Special | 90.954 | Tommy Milton | 108.17 |
| 1924 | L. L. Corum & Joe Boyer (21) | Duesenberg Special | 98.234 | Jimmy Murphy | 108.037 |
| 1925 | Peter DePaolo (2) | Duesenberg Special | 101.127 | Leon Duray | 113.196 |
| 1926 | Frank Lockhart (20) | Miller Special | 95.904 | Earl Cooper | 111.735 |
| 1927 | George Souders (22) | Duesenberg | 97.545 | Frank Lockhart | 120.100 |
| 1928 | Louie Meyer (13) | Miller Special | 99.482 | Leon Duray | 122.391 |
| 1929 | Ray Keech (6) | Simplex Piston Ring Special | 97.585 | Cliff Woodbury | 120.599 |
| 1930 | Billy Arnold (1) | Miller-Hartz Special | 100.448 | Billy Arnold | 113.268 |
| 1931 | Louis Schneider (13) | Bowes Seal Fast Special | 96.629 | Russ Snowberger | 112.796 |
| 1932 | Fred Frame (27) | Miller-Hartz Special | 104.144 | Lou Moore | 117.363 |
| 1933 | Louie Meyer (6) | Tydol Special | 104.162 | Bill Cummings | 118.530 |
| 1934 | Bill Cummings (10) | Boyle Products Special | 104.863 | Kelly Petillo | 119.329 |
| 1935 | Kelly Petillo (22) | Gilmore Speedway Special | 106.240 | Rex Mays | 120.736 |
| 1936 | Louie Meyer (28) | Ring Free Special | 109.069 | Rex Mays | 119.644 |
| 1937 | Wilbur Shaw (2) | Shaw-Gilmore Special | 113.580 | Bill Cummings | 123.343 |
| 1938 | Floyd Roberts (1) | Burd Piston Ring Special | 117.200 | Floyd Roberts | 125.681 |
| 1939 | Wilbur Shaw (3) | Boyle Special | 115.035 | Jimmy Snyder | 130.138 |
| 1940 | Wilbur Shaw (2) | Boyle Special | 114.277 | Rex Mays | 127.850 |

RacingMath

# Indianapolis 500 statistics

| YEAR | WINNER (position) | CAR | MPH | POLE SITTER | MPH |
|---|---|---|---|---|---|
| 1941 | Floyd Davis & Mauri Rose (17) | Noc-Out Hose Clamp Special | 115.117 | Mauri Rose | 128.691 |
| 1942–5 | Not held World War II | | | | |
| 1946 | George Robson (15) | Thorne Engineering Special | 114.820 | Cliff Bergere | 126.471 |
| 1947 | Mauri Rose (3) | Blue Crown Spark Plug Special | 116.338 | Ted Horn | 126.564 |
| 1948 | Mauri Rose (3) | Blue Crown Spark Plug Special | 119.814 | Duke Nalon | 131.603 |
| 1949 | Bill Holland (4) | Blue Crown Spark Plug Special | 121.327 | Duke Nalon | 132.939 |
| 1950 | Johnnie Parsons (5) | Wynn's Friction Proofing | 124.002 | Walt Faulkner | 134.343 |
| 1951 | Lee Wallard (2) | Belanger Special | 126.244 | Duke Nalon | 136.498 |
| 1952 | Troy Ruttman (7) | Agajanian Special | 128.922 | Fred Agabashian | 138.010 |
| 1953 | Bill Vukovich (1) | Fuel Injection Special | 128.740 | Bill Vukovich | 138.392 |
| 1954 | Bill Vukovich (19) | Fuel Injection Special | 130.840 | Jack McGrath | 141.033 |
| 1955 | Bob Sweikert (14) | John Zink Special | 128.213 | Jerry Hoyt | 140.045 |
| 1956 | Pat Flaherty (1) | John Zink Special | 128.490 | Pat Flaherty | 145.596 |
| 1957 | Sam Hanks (13) | Belond Exhaust Special | 135.601 | Pat O'Connor | 143.948 |
| 1958 | Jimmy Bryan (7) | Belond AP Parts Special | 133.791 | Dick Rathmann | 145.974 |
| 1959 | Rodger Ward (6) | Leader Card 500 Roadster | 135.857 | Johnny Thomson | 145.908 |
| 1960 | Jim Rathmann (2) | Ken-Paul Special | 138.767 | Eddie Sachs | 146.592 |
| 1961 | A. J. Foyt (7) | Bowes Seal Fast Special | 139.130 | Eddie Sachs | 147.481 |
| 1962 | Rodger Ward (2) | Leader Card 500 Roadster | 140.293 | Parnelli Jones | 150.370 |
| 1963 | Parnelli Jones (1) | Agajanian-Willard Special | 143.137 | Parnelli Jones | 151.153 |
| 1964 | A. J. Foyt (5) | Sheraton-Thompson Special | 147.350 | Jim Clark | 158.828 |
| 1965 | Jim Clark (2) | Lotus Ford | 150.686 | A. J. Foyt | 161.233 |
| 1966 | Graham Hill (15) | American Red Ball Special | 144.317 | Mario Andretti | 165.899 |
| 1967 | A. J. Foyt (4) | Sheraton-Thompson Special | 151.207 | Mario Andretti | 168.982 |
| 1968 | Bobby Unser (3) | Rislone Special | 152.882 | Joe Leonard | 171.559 |
| 1969 | Mario Andretti (2) | STP Oil Treatment Special | 156.867 | A. J. Foyt | 170.568 |
| 1970 | Al Unser (1) | Johnny Lightning Special | 155.749 | Al Unser | 170.221 |
| 1971 | Al Unser (5) | Johnny Lightning Special | 157.735 | Peter Revson | 178.696 |
| 1972 | Mark Donohue (3) | Sunoco McLaren | 162.962 | Bobby Unser | 195.940 |

## Indianapolis 500 statistics

| YEAR | WINNER (position) | CAR | MPH | POLE SITTER | MPH |
|---|---|---|---|---|---|
| 1973 | Gordon Johncock (11) | STP Double Oil Filters | 159.036 | A. J. Foyt | 198.413 |
| 1974 | Johnny Rutherford (25) | McLaren | 158.589 | Johnny Rutherford | 191.632 |
| 1975 | Bobby Unser (3) | Jorgensen Eagle | 49.213 | A. J. Foyt | 193.976 |
| 1976 | Johnny Rutherford (1) | Hy-Gain McLaren/Goodyear | 148.725 | Johnny Rutherford | 188.957 |
| 1977 | A. J. Foyt (4) | Gilmore Racing Team | 161.331 | Tom Sneva | 198.884 |
| 1978 | Al Unser (5) | FNCTC Chaparral Lola | 161.363 | Tom Sneva | 202.156 |
| 1979 | Rick Mears (1) | The Gould Charge | 158.899 | Rick Mears | 193.736 |
| 1980 | Johnny Rutherford (1) | Pennzoil Chaparral | 142.862 | Johnny Rutherford | 192.256 |
| 1981 | Bobby Unser (1) | Norton Spirit Penske PC-9B | 139.084 | Bobby Unser | 200.546 |
| 1982 | Gordon Johncock (5) | STP Oil Treatment | 162.029 | Rick Mears | 207.004 |
| 1983 | Tom Sneva (4) | Texaco Star | 162.117 | Teo Fabi | 207.395 |
| 1984 | Rick Mears (3) | Pennzoil Z-7 | 163.612 | Tom Sneva | 210.029 |
| 1985 | Danny Sullivan (8) | Miller American Special | 152.982 | Pancho Carter | 212.583 |
| 1986 | Bobby Rahal (4) | Budweiser/Truesports/March | 170.722 | Rick Mears | 216.828 |
| 1987 | Al Unser (20) | Cummins Holset Turbo | 162.175 | Mario Andretti | 215.390 |
| 1988 | Rick Mears (1) | Pennzoil Z-7/Penske Chevy V-8 | 144.809 | Rick Mears | 219.198 |
| 1989 | Emerson Fittipaldi (3) | Marlboro/Penske Chevy V-8 | 167.581 | Rick Mears | 223.885 |
| 1990 | Arie Luyendyk (3) | Domino's Pizza Chevrolet | 185.98 | Emerson Fittipaldi | 225.301 |
| 1991 | Rick Mears (1) | Marlboro Penske Chevy | 176.457 | Rick Mears | 224.113 |
| 1992 | Al Unser, Jr. (12) | Valvoline Galmer '92 | 134.477 | Roberto Guerrero | 232.482 |
| 1993 | Emerson Fittipaldi (9) | Marlboro Penske Chevy | 157.207 | Arie Luyendyk | 223.967 |
| 1994 | Al Unser, Jr. (1) | Marlboro Penske Mercedes | 160.872 | Al Unser, Jr. | 228.011 |
| 1995 | Jacques Villeneuve (5) | Player's Ltd. Reynard Ford | 153.616 | Scott Brayton | 231.604 |
| 1996 | Buddy Lazier (5) | Reynard Ford | 147.066 | Tony Stewart | 233.100 |
| 1997 | | | | | |
| 1998 | | | | | |
| 1999 | | | | | |
| 2000 | | | | | |

From *The 1996 Information Please® Sports Almanac*. Copyright © 1995 by Information Please LLC. All rights reserved. Reprinted by permission of INSO™ Corporation.

RacingMath

# Engine blocks and blockheads

When it comes to car engines, do you know all about displacement, bore times stroke, and horsepower? Or are you a complete blockhead when it comes to engine blocks?

Study the information below. On a separate sheet of paper, write a short paragraph for each item. Explain what the information means. Define any terms. Explain the importance of these things in a racing car.

Finally, find the owner's manual for your family car, or for a neighbor's or friend's car. Make a chart listing similar information for that car.

### Sports car

| | |
|---|---|
| 1. Engine type | aluminum block and heads, V-12 |
| 2. Displacement | 348 cu. in./5707 cc |
| 3. Bore x stroke | 3.43 x 3.15 in./87.0 x 80.0 mm |
| 4. Compression ratio | 10.0:1 |
| 5. Horsepower | 490 hp @ 5200 rpm |
| 6. Maximum engine speed | 7600 rpm |

# Hot rod season

The chart that follows gives the results of the 1995 National Hot Rod Association Drag Racing championships in 3 categories: Top Fuel, Funny Car, and Pro Stock.

All times are based on 2 cars racing head-to-head from a standing start. The course is always a quarter-mile long straight.

1. Explain what the time numbers indicate: hours? minutes? seconds? How do you know what they mean?

2. For each of the 3 categories, what was the fastest winning speed? What was the slowest winning speed? the average (mean) winning speed? the median winning speed?

3. For each of the 3 categories, what was the fastest second-place speed? What was the slowest second-place speed? the average (mean) second-place speed? the median second-place speed?

4. Which of the 3 categories has the fastest winning speeds? Which has the slowest? What is the difference in miles per hour between the fastest and the slowest categories?

5. Study the results of the 1995 Hot Rod races carefully. How many instances can you find in which the second-place driver drove faster than the first-place driver? List each such incident.

6. How is this possible? You may have to do some research on drag racing to learn the answer.

7. Research the Top Fuel, Funny Car, and Pro Stock categories. Write a 2-sentence description of each category.

RacingMath

## 1995 season hot rod drag racing

| DATE | EVENT | WINNER | TIME | MPH | 2ND PLACE | TIME | MPH |
|---|---|---|---|---|---|---|---|
| Feb. 5 | Winter Nationals | | | | | | |
| | Top Fuel | Eddie Hill | 4.859 | 299.50 | S. Kalitta | 11.879 | 77.31 |
| | Funny Car | Cruz Pedregon | 5.304 | 278.72 | C. Etchells | broke | |
| | Pro Stock | Darrell Alderman | 7.054 | 196.03 | S. Geoffrion | 7.105 | 195.18 |
| Feb. 19 | ATSCO Nationals | | | | | | |
| | Top Fuel | Larry Dixon | 4.821 | 300.00 | S. Anderson | 8.378 | 89.48 |
| | Funny Car | John Force | 5.057 | 298.30 | T. Hoover | 5.404 | 238.28 |
| | Pro Stock | Darrell Alderman | 7.073 | 194.46 | J. Eckman | 7.626 | 144.92 |
| Mar. 12 | Slick 50 Nationals | | | | | | |
| | Top Fuel | Mike Dunn | 4.857 | 296.34 | K. Bernstein | 8.292 | 87.67 |
| | Funny Car | Al Hofmann | 5.207 | 293.15 | M. Oswald | 5.315 | 284.53 |
| | Pro Stock | Scott Geoffrion | 7.062 | 96.03 | J. Yates | 7.096 | 195.22 |
| Apr. 2 | Gatornationals | | | | | | |
| | Top Fuel | Connie Kalitta | 4.794 | 290.79 | S. Kalitta | 4.954 | 271.65 |
| | Funny Car | John Force | 5.347 | 263.00 | K. C. Spurlock | 6.372 | 149.80 |
| | Pro Stock | Warren Johnson | 7.136 | 195.14 | D. Alderman | 7.119* | 195.27 |
| Apr. 24 | Fram Nationals | | | | | | |
| | Top Fuel | Cory McClenathan | 4.806 | 298.90 | R. Capps | 5.079 | 236.34 |
| | Funny Car | John Force | 5.174 | 297.22 | C. Pedregon | 13.321 | 73.14 |
| | Pro Stock | Mark Osborne | 7.141 | 194.17 | D. Alderman | broke | |
| May 7 | Mid-South Nationals | | | | | | |
| | Top Fuel | Cory McClenathan | 4.810 | 307.48 | M. Dunn | 4.997 | 288.27 |
| | Funny Car | Gary Clapshaw | 5.339 | 286.89 | G. Densham | 5.457 | 276.66 |
| | Pro Stock | Mark Pawuk | 7.195 | 192.80 | J. Eckman | 7.213 | 192.84 |
| May 21 | Mopar Nationals | | | | | | |
| | Top Fuel | Larry Dixon | 4.991 | 281.77 | S. Kalitta | 5.020 | 290.79 |
| | Funny Car | Cruz Pedregon | 5.246 | 294.31 | D. Skuza | 7.407 | 121.11 |
| | Pro Stock | Bob Glidden | 7.117 | 194.42 | J. Yates | 7.123 | 194.67 |

*False start

# 1995 season hot rod drag racing

| DATE | EVENT | | WINNER | TIME | MPH | 2ND PLACE | TIME | MPH |
|---|---|---|---|---|---|---|---|---|
| June 4 | Virginia Nationals | | | | | | | |
| | | Top Fuel | Cory McClenathan | 4.962 | 293.82 | B. Johnson | 4.980 | 282.39 |
| | | Funny Car | John Force | 5.193 | 270.59 | K. C. Spurlock | 5.570 | 246.23 |
| | | Pro Stock | Warren Johnson | 7.062 | 196.03 | J. Yates | 7.190 | 195.18 |
| June 12 | Spring Nationals | | | | | | | |
| | | Top Fuel | Scott Kalitta | 4.772 | 305.18 | E. Hill | 4.884 | 287.53 |
| | | Funny Car | Al Hofmann | 5.125 | 299.90 | J. Force | 9.542 | 85.39 |
| | | Pro Stock | Steve Schmidt | 7.125 | 193.21 | W. Johnson | 7.121 | 192.47 |
| July 2 | West. Auto Nationals | | | | | | | |
| | | Top Fuel | Scott Kalitta | 4.820 | 295.27 | K. Bernstein | 9.548 | 88.85 |
| | | Funny Car | Cruz Pedregon | 5.192 | 273.39 | K.C. Spurlock | 9.434 | 98.25 |
| | | Pro Stock | Warren Johnson | 7.069 | 194.46 | J. Yates | 7.128 | 193.54 |
| July 23 | Mile-High Nationals | | | | | | | |
| | | Top Fuel | Scott Kalitta | 4.813 | 298.30 | B. Johnson | 5.022 | 261.17 |
| | | Funny Car | John Force | 5.258 | 291.16 | A. Hofmann | 10.175 | 82.36 |
| | | Pro Stock | Kurt Johnson | 7.491 | 183.29 | J. Yates | 7.465 | 183.00 |
| July 30 | Autolight Nationals | | | | | | | |
| | | Top Fuel | Mike Dunn | 5.107 | 276.83 | P. Austin | 6.432 | 141.37 |
| | | Funny Car | Al Hofmann | 5.302 | 281.07 | J. Force | 5.330 | 286.44 |
| | | Pro Stock | Jim Yates | 7.143 | 193.79 | W. Johnson | 7.138 | 194.55 |
| Aug. 8 | Northwest Nationals | | | | | | | |
| | | Top Fuel | Ron Capps | 4.930 | 295.76 | C. McClenathan | 12.367 | 77.64 |
| | | Funny Car | Al Hofmann | 5.200 | 285.80 | G. Densham | 5.299 | 279.93 |
| | | Pro Stock | Warren Johnson | 7.022 | 197.23 | B. Glidden | 7.056 | 195.73 |
| Aug. 20 | Champion Nationals | | | | | | | |
| | | Top Fuel | Mike Dunn | 4.952 | 291.63 | T. Johnson Jr. | 6.243 | 160.68 |
| | | Funny Car | John Force | 5.284 | 291.92 | G. Densham | 5.850 | 231.06 |
| | | Pro Stock | Warren Johnson | 7.202 | 193.05 | J. Yates | 7.194 | 191.24 |

RacingMath

## 1995 season hot rod drag racing

| DATE | EVENT | WINNER | TIME | MPH | 2ND PLACE | TIME | MPH |
|---|---|---|---|---|---|---|---|
| Sept. 4 | Keystone Nationals | | | | | | |
| | Top Fuel | Larry Dixon | 4.931 | 293.25 | B. Vendorgriff Jr. | 5.330 | 181.59 |
| | Funny Car | Cruz Pedregon | 5.075 | 304.67 | J. Force | 5.228 | 290.60 |
| | Pro Stock | Warren Johnson | 7.059 | 197.02 | L. Warden | 7.126 | 192.51 |
| Sept. 17 | Keystone Nationals | | | | | | |
| | Top Fuel | Scott Kalitta | 4.801 | 298.90 | L. Dixon | 5.937 | 150.88 |
| | Funny Car | Chuck Etchells | 6.121 | 264.31 | J. Force | 6.350 | 242.19 |
| | Pro Stock | Warren Johnson | 7.060 | 195.43 | M. Pawuk | 8.980 | 101.63 |
| Oct. 1 | Sears Nationals | | | | | | |
| | Top Fuel | Scott Kalitta | 4.756 | 295.46 | M. Dunn | 5.143 | 214.43 |
| | Funny Car | Cruz Pedregon | 5.068 | 301.81 | A. Hofmann | 5.148 | 296.05 |
| | Pro Stock | Steve Schmidt | 7.076 | 194.67 | J. Eckman | 10.867 | 83.36 |

From *The 1996 Information Please® Sports Almanac.* Copyright © 1995 by Information Please LLC. All rights reserved. Reprinted by permission of INSO™ Corporation.

RacingMath 84

# Autocross racing

In autocross racing, drivers compete against a time clock. The one who can negotiate a course laid out with traffic cones in the fastest time, without hitting any of the cones, wins that autocross category race.

**1.** Research autocross racing and write a brief history of it. Include when and where it started, other names it has gone by in the past, and so on.

**2.** Autocross events are laid out on airstrips, in parking malls, or even on roadways, with traffic cones marking the course. Traffic cones are also called pylons. What is the origin of the word *pylon*? Write a brief description of other kinds of pylons.

**3.** Imagine 2 spectators at an autocross event at the local mall. Write a comic skit about the event from the point of view of the spectators.

**4.** Call an auto supply store or hardware store and ask if they sell pylons. Find a store that sells the orange plastic pylons used in autocross (or the smaller pylons used in rollerblade racing). Go there and measure the base of the pylon. What are the dimensions of the pylon's base? (If the store sells more than one size pylon, record both.)

**RacingMath**

**5.** What is the diameter of the circle at the base of the pylon? (This is the circle you see when you turn the pylon upside-down and look inside.)

**6.** What is the height of the pylon?

**7.** Stack one pylon on top of another and measure the combined height. Then stack a third pylon on top and measure that height. What is the average increase in height each time a pylon is added to the column?

**8.** How tall would a column of 10 pylons be? a column of 50 pylons?

**9.** If possible, go to an autocross event and observe several races. What kind of skills do the drivers need? Do you think these skills are important? Why?

**10.** Design a newsletter or stationery for an autocross racer. Give the newsletter a name and a logo. If you choose to design stationery, give the driver a name, an address, and a logo.

### Challenge problem

This is a tough one. Learn the formula for finding the volume of a cone. What is the volume of the cone portion of the pylon you measured?

# Hot rods and funny cars

The 2 tables below show hot rod records through 1994. All records pertain to speed on a straight quarter-mile raceway, or drag.

### Pro stock category

| MPH | Driver | Date record set |
|---|---|---|
| 197.80 | Darrell Alderman | 7/30/94 |
| 197.49 | Warren Johnson | 4/8/94 |
| 197.15 | Warren Johnson | 4/23/94 |
| 196.97 | Warren Johnson | 4/8/94 |
| 196.93 | Scott Geoffrion | 5/20/94 |

From *The 1996 Information Please® Almanac.* Copyright © 1995 by Information Please LLC. All rights reserved. Reprinted by permission of INSO™ Corporation.

### Funny car category

| MPH | Driver | Date record set |
|---|---|---|
| 303.95 | John Force | 10/30/94 |
| 303.74 | John Force | 10/30/94 |
| 302.82 | John Force | 10/29/94 |
| 302.11 | John Force | 9/18/94 |
| 301.40 | John Force | 10/28/94 |
| 301.40 | John Force | 10/29/94 |

From *The 1996 Information Please® Almanac.* Copyright© 1995 by Information Please LLC. All rights reserved. Reprinted by permission of INSO™ Corporation.

**1.** Be a mathematical detective. Write a list of 5 conclusions you can draw about hot rod racing from the two charts above. Remember: These are records of the fastest times in hot rod racing from its beginning through the end of 1994.

**2.** Have the records in these 2 charts been broken? Check a reference source such as an almanac and create the latest records chart for each category.

**3.** Write a story about a funny man with funny money in a funny car.

RacingMath

# Le Mans 24-hour statistics

Look at the chart that follows to learn something about the Le Mans 24-Hour Race. Then complete the projects below.

1. First, bring the Le Mans 24-Hour statistics up-to-date by filling in the information for each year from 1997 to the present.

2. Then determine the average winning speed in the Le Mans 24-Hour Race decade by decade. Use 1923 to 1929 as one decade labeled 1920s, use 1930 to 1939 as a decade labeled 1930s, and so on. Make a chart showing the average winning speed.

3. Study the decade-by-decade statistics. Make a prediction about the average winning speed in the Le Mans 24-Hour Race for the next decade.

4. Which drivers have won the Le Mans race more than once? Make a brief list of their names and the number of times each has won the Le Mans. Write a poem about the multiple winners.

5. Read about the history of the 24 Hours of Le Mans. Write an encyclopedia article of 150 words or less explaining the race to somebody who knows nothing about it.

6. Up until 1970, the Le Mans race was started as drivers raced across the track to their cars, jumped into the cars, started the engines, and raced away. Can you find any photographs or paintings depicting these world-famous starts?

7. Do some historical research to discover why the running start was discontinued. Write a report on the subject.

8. The Le Mans race is run by teams, with 2 or 3 drivers on each. Why? Read about this and write a report on how it works.

9. Find a diagram of the present-day Le Mans course. Trace it or photocopy it. Label the direction cars race and name any famous parts of the course.

10. Make a chart showing how many times each car manufacturer has won Le Mans. For purposes of the chart, disregard the model numbers of the cars: a Porsche is a Porsche, whether it's a 936, a 956, or a 962. Organize the chart so that the car that has won most often is at the top, the second-place winner is below that, and so on.

# The 24 hours of Le Mans

| YEAR | DRIVERS | CAR | MPH |
|---|---|---|---|
| 1923 | Andre Lagache & Rene Leonard | Chenard & Walcker | 57.21 |
| 1924 | John Duff & Francis Clement | Bentley | 53.78 |
| 1925 | Gerard de Courcelles & Andre Rossignol | La Lorraine | 57.84 |
| 1926 | Robert Bloch & Andre Rossignol | La Lorraine | 66.08 |
| 1927 | J.D. Benjafield & Sammy Davis | Bentley | 61.35 |
| 1928 | Woolf Barnato & Bernard Rubin | Bentley | 69.11 |
| 1929 | Woolf Barnato & Sir Henry Birkin | Bentley Speed 6 | 73.63 |
| 1930 | Woolf Barnato & Glen Kidston | Bentley Speed 6 | 75.88 |
| 1931 | Earl Howe & Sir Henry Birkin | Alfa Romeo | 78.13 |
| 1932 | Raymond Sommer & Luigi Chinetti | Alfa Romeo | 76.48 |
| 1933 | Raymond Sommer & Tazio Nuvolari | Alfa Romeo | 81.40 |
| 1934 | Luigi Chinetti & Philippe Etancelin | Alfa Romeo | 74.74 |
| 1935 | John Hindmarsh & Louis Fontes | Lagonda | 77.85 |
| 1936 | Not held | | |
| 1937 | Jean-Pierre Wimille & Robert Benoist | Bugatti 57G | 85.13 |
| 1938 | Eugene Chaboud & Jean Tremoulet | Delahaye | 82.36 |
| 1939 | Jean-Pierre Wimille & Pierre Veyron | Bugatti 57G | 86.86 |
| 1940–48 | Not held World War II | | |
| 1949 | Luigi Chinetti & Lord Selsdon | Ferrari | 82.28 |
| 1950 | Louis Rosier & Jean-Louis Rosier | Talbot-Lago | 89.71 |
| 1951 | Peter Walker & Peter Whitehead | Jaguar C | 93.50 |
| 1952 | Hermann Lang & Fritz Reiss | Mercedes-Benz | 96.67 |
| 1953 | Tony Rolt & Duncan Hamilton | Jaguar C | 98.65 |
| 1954 | Froilan Gonzalez & Maurice Trintignant | Ferrari 375 | 105.13 |
| 1955 | Mike Hawthorn & Ivor Bueb | Jaguar D | 107.05 |
| 1956 | Ron Flockhart & Ninian Sanderson | Jaguar D | 104.47 |
| 1957 | Ron Flockhart & Ivor Bueb | Jaguar D | 113.83 |
| 1958 | Olivier Gendebien & Phil Hill | Ferrari 250 | 106.18 |
| 1959 | Roy Salvadori & Carroll Shelby | Aston Martin | 112.55 |

# The 24 hours of Le Mans

| YEAR | DRIVERS | CAR | MPH |
|---|---|---|---|
| 1960 | Olivier Gendebien & Paul Friere | Ferrari 250 | 109.17 |
| 1961 | Olivier Gendebien & Phil Hill | Ferrari 250 | 115.88 |
| 1962 | Olivier Gendebien & Phil Hill | Ferrari 250 | 115.22 |
| 1963 | Lodovico Scarfiotti & Lorenzo Bandini | Ferrari 250 | 118.08 |
| 1964 | Jean Guichel & Nino Vaccarella | Ferrari 275 | 121.54 |
| 1965 | Masten Gregory & Jochen Rindt | Ferrari 250 | 121.07 |
| 1966 | Bruce McLaren & Chris Amon | Ford Mk. II | 125.37 |
| 1967 | A. J. Foyt & Dan Gurney | Ford Mk. IV | 135.46 |
| 1968 | Pedro Rodriguez & Lucien Bianchi | Ford GT40 | 115.27 |
| 1969 | Jacky Ickx & Jackie Oliver | Ford GT40 | 129.38 |
| 1970 | Hans Herrmann & Richard Attwood | Porsche 917 | 119.28 |
| 1971 | Gijs van Lennep & Helmut Marko | Porsche 917 | 138.13 |
| 1972 | Graham Hill & Henri Pescarolo | Matra-Simca | 121.45 |
| 1973 | Henri Pescarolo & Gerard Larrousse | Matra-Simca | 125.67 |
| 1974 | Henri Pescarolo & Gerard Larrousse | Matra-Simca | 119.27 |
| 1975 | Derek Bell & Jacky Ickx | Mirage-Ford | 118.98 |
| 1976 | Jacky Ickx & Gijs van Lennep | Porsche 936 | 123.49 |
| 1977 | Jacky Ickx, Jurgen Barth & Hurley Haywood | Porsche 936 | 120.95 |
| 1978 | Jean-Pierre Jaussaud & Didier Pironi | Renault-Alpine | 130.60 |
| 1979 | Klaus Ludwig, Bill Wittington & Don Whittington | Porsche 935 | 108.10 |
| 1980 | Jean-Pierre Jaussaud & Jean Rondeau | Rondeau-Cosworth | 119.23 |
| 1881 | Jacky Ickx & Derek Bell | Porsche 936 | 124.94 |
| 1982 | Jacky Ickx & Derek Bell | Porsche 956 | 126.85 |
| 1983 | Vern Schuppan, Hurley Haywood & Al Holbert | Porsche 956 | 130.70 |
| 1984 | Klaus Ludwig & Henri Pescarolo | Porsche 956 | 126.88 |
| 1985 | Klaus Ludwig, Paolo Barilla & John Winter | Porsche 956 | 131.75 |
| 1986 | Derek Bell, Hans Stuck & Al Holbert | Porsche 962 | 128.75 |
| 1987 | Derek Bell, Hans Stuck & Al Holbert | Porsche 962 | 124.06 |
| 1988 | Jan Lammers, Johnny Dumfries & Andy Wallace | Jaguar XJR | 137.75 |

## The 24 hours of Le Mans

| YEAR | DRIVERS | CAR | MPH |
|---|---|---|---|
| 1989 | Jochen Mass, Manuel Reuter & Stanley Dickens | Sauber-Mercedes | 136.39 |
| 1990 | John Nielsen, Price Cobb & Martin Brundle | Jaguar XJR-12 | 126.71 |
| 1991 | Volker Weider, Johnny Herbert & Bertrand Gachof | Mazda 787B | 127.31 |
| 1992 | Derek Warwick, Yannick Dalmas & Mark Blundell | Peugeot 905B | 123.89 |
| 1993 | Geoff Brabham, Christophe Bouchut & Eric Helary | Peugeot 905 | 132.58 |
| 1994 | Yannick Dalmas, Hurley Haywood & Mauro Baldi | Porsche 962LM | 129.82 |
| 1995 | Yannick Dalmas, J. J. Lehto & Masanori Sekiya | McLaren BMW | 105.00 |
| 1996 | Davy Jones, Manuel Reuter, & Alexander Wurz | TWR Porsche | 124.65 |
| 1997 | | | |
| 1998 | | | |
| 1999 | | | |
| 2000 | | | |

From *The 1996* Information *Please®* Almanac. Copyright © 1995 by Information Please LLC. All rights reserved. Reprinted by permission of INSO™ Corporation.

RacingMath

# Pistons and power

For the projects on this page, you should visit your local library or consult an auto mechanic who's willing to spend some time answering your questions and showing you how engines work.

1. Explain what cylinders and pistons are and how they work.

2. Car magazines talk about bore and stroke dimensions when discussing engine features and power. What is the bore? What is the stroke dimension?

3. Define *horsepower*.

4. How are bore and stroke dimensions related to the power output of an engine? Give examples.

5. How is the number of cylinders related to the power output of an engine?

6. How can you increase the power of an automobile engine? List several ways. Discuss the positive and negative effects of each method.

7. Give an example of how modifying the bore and stroke dimensions can increase the power output of an engine. In your example, give a bore and stroke size and power output. Show what happens when the dimensions are increased.

# Daytona 500 NASCAR statistics

These charts give statistics for the Daytona 500 NASCAR race. Study the charts and complete the projects that follow.

## Daytona 500 statistics

| YEAR | WINNER | CAR | MPH | POLE SITTER | MPH |
|---|---|---|---|---|---|
| 1959 | Lee Petty | Oldsmobile | 135.521 | Bob Welborn | 140.121 |
| 1960 | Junior Johnson | Chevrolet | 124.740 | Cotton Owens | 149.892 |
| 1961 | Marvin Panch | Pontiac | 149.601 | Fireball Roberts | 155.709 |
| 1962 | Fireball Roberts | Pontiac | 152.529 | Fireball Roberts | 156.999 |
| 1963 | Tiny Lund | Ford | 151.566 | Fireball Roberts | 160.943 |
| 1964 | Richard Petty | Plymouth | 154.334 | Paul Goldsmith | 174.910 |
| 1965 | Fred Lorenzen | Ford | 141.539 | Darel Dieringer | 171.151 |
| 1966 | Richard Petty | Plymouth | 160.627 | Richard Petty | 175.165 |
| 1967 | Mario Andretti | Ford | 149.926 | Curtis Turner | 180.831 |
| 1968 | Cale Yarborough | Mercury | 143.251 | Cale Yarborough | 189.222 |
| 1969 | Lee Roy Yarbrough | Ford | 157.950 | Buddy Baker | 188.901 |
| 1970 | Pete Hamilton | Plymouth | 149.601 | Cale Yarborough | 194.015 |
| 1971 | Richard Petty | Plymouth | 144.462 | A. J. Foyt | 182.744 |
| 1972 | A. J. Foyt | Mercury | 161.550 | Bobby Issac | 186.632 |
| 1973 | Richard Petty | Dodge | 157.205 | Buddy Baker | 185.662 |
| 1974 | Richard Petty | Dodge | 140.894 | David Pearson | 185.017 |
| 1975 | Benny Parsons | Chevrolet | 153.649 | Donnie Allison | 185.827 |
| 1976 | David Pearson | Mercury | 152.181 | Ramo Stott | 183.456 |
| 1977 | Cale Yarborough | Chevrolet | 153.218 | Donnie Allison | 188.048 |

RacingMath

## Daytona 500 statistics

| YEAR | WINNER | CAR | MPH | POLE SITTER | MPH |
|---|---|---|---|---|---|
| 1978 | Bobby Allison | Ford | 159.730 | Cale Yarborough | 187.536 |
| 1979 | Richard Petty | Oldsmobile | 143.977 | Buddy Baker | 196.049 |
| 1980 | Buddy Baker | Oldsmobile | 177.602 | Buddy Baker | 194.099 |
| 1981 | Richard Petty | Buick | 169.651 | Bobby Allison | 194.624 |
| 1982 | Bobby Allison | Buick | 153.991 | Benny Parsons | 196.317 |
| 1983 | Cale Yarborough | Pontiac | 155.979 | Ricky Rudd | 198.864 |
| 1984 | Cale Yarborough | Chevrolet | 150.994 | Cale Yarborough | 201.848 |
| 1985 | Bill Elliott | Ford | 172.265 | Bill Elliott | 205.114 |
| 1986 | Geoff Bodine | Chevrolet | 148.124 | Bill Elliott | 205.039 |
| 1987 | Bill Elliott | Ford | 176.263 | Bill Elliott | 210.364 |
| 1988 | Bobby Allison | Buick | 137.531 | Ken Schrader | 198.823 |
| 1989 | Darrell Waltrip | Chevrolet | 148.466 | Ken Schrader | 196.996 |
| 1990 | Derrike Cope | Chevrolet | 165.761 | Ken Schrader | 196.515 |
| 1991 | Ernie Irvan | Chevrolet | 148.148 | Davey Allison | 195.955 |
| 1992 | Davey Allison | Ford | 160.256 | Sterling Martin | 192.213 |
| 1993 | Dale Jarrett | Chevrolet | 154.972 | Kyle Petty | 189.426 |
| 1994 | Sterling Marlin | Chevrolet | 156.931 | Loy Allen | 190.158 |
| 1995 | Sterling Marlin | Chevrolet | 141.710 | Dale Jarrett | 193.498 |
| 1996 | Dale Jarrett | Ford | 154.308 | Dale Earnhardt | 189.510 |
| 1997 | | | | | |
| 1998 | | | | | |
| 1999 | | | | | |
| 2000 | | | | | |

From *The 1996 Information Please® Almanac.* Copyright © 1995 by Information Please LLC. All rights reserved. Reprinted by permission of INSO™ Corporation.

1. First, bring the Daytona 500 statistics up-to-date by filling in the information for each year from 1997 to the present.

2. Then determine the average winning speed in the Daytona 500, decade by decade. Use 1959 as a single decade labeled 1950s, use 1960 to 1969 as a decade labeled 1960s, and so on. Make a chart showing the average winning speed. Leave space on the chart for the project.

3. What was the average pole position speed decade by decade? Put it on the chart you made.

4. In what percent of cases did the Daytona 500 pole sitter win the race?

5. Which drivers have won the Daytona 500 more than once? Choose 1 of these drivers and write a 300-word biography of him.

6. Draw or photocopy the Daytona 500 track. Compare it to the Le Mans 24-Hour Race course. List 5 differences in the tracks. After each difference, explain how that difference affects results.

7. Access NASCAR information online. Download any Daytona 500 information. What do online sources emphasize and why: numbers? excitement? names of drivers? types of cars?

### Challenge problem

Compare the Daytona 500 decade-by-decade speeds to those for the Indianapolis 500 and the Le Mans 24-Hour Race. Make a chart comparing the speed of the three world-famous races, decade by decade.

What conclusions do you draw from studying the differences between the three races in the average decade-by-decade speed?

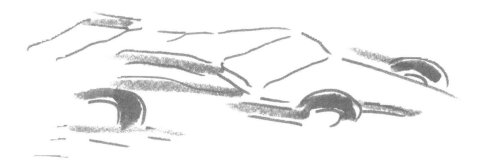

# Pikes Peak or bust

In your school or town library, conduct some research on Pikes Peak and the Pikes Peak Hill Climb. You can do this alone or with the help of 2 or 3 classmates. Answer each of the questions below and on page 97.

1. Which European American is credited with discovering Pikes Peak? When?

2. Why is the mountain called Pikes Peak?

3. Which tribes of Native Americans lived in the Pikes Peak region? Did/does the mountain have any special significance for them?

4. When was the Pikes Peak Hill Climb started? by whom?

5. How many classes of vehicles participated in the first race? How many classes of vehicles participate today? List them.

6. Who won the first Pikes Peak Hill Climb, and what was his speed? Who were the winners of the most recent climb, and what were their speeds?

7. What is the height of Pikes Peak? Make a chart listing the 5 highest peaks in the United States. For each peak, give its name, location, height, and any special information about it.

8. Make a chart listing the 5 highest peaks in the world. For each peak, give its name, location, height, and any special information about it. Are any of the mountains on the U.S. list also on the world list?

**9.** What is the median altitude of the 5 U.S. peaks? Of the 5 world peaks? What is the difference in median altitude between the 2 groups? What is the average (mean) altitude of the 5 U.S. mountains? of the 5 world mountains? What is the difference in average altitude between the 2 groups?

**10.** Design a Pikes Peak Hill Climb cap, T-shirt, jacket, or Web page. The design should "say" something, about or hint at both the mountain and the auto race.

### Challenge problem

On a map of North America, find the northernmost and southernmost tips of the Rocky Mountains. Measure the distance between them in miles or kilometers. Measure the chain at its widest part. In square miles or square kilometers, what is the area of the Rocky Mountains? How does this compare to the area of individual states in the United States or of individual provinces in Canada?

# Fabulous Ferraris

For 1997 Ferrari introduced 4 new cars: the 550 Maranello, the 456 GTA, the F355 Spyder, and the F50. The 550 Maranello is a front-engined, rear-drive, 2-seater touring car. The 456 GTA has an automatic transmission and room for 4, as long as those sitting in the back seat are short. The F355 Spyder is a 2-seater convertible; and the F50, derived from Ferrari's Formula One car, is the closest thing to a rocketship on wheels.

The chart below gives you factual information on the four new Ferrari models. After studying the chart, you will be able to answer the questions that follow.

|  | 550 Maranello | F355 GTA | 456 Spyder | F50 |
| --- | --- | --- | --- | --- |
| List Price | $200,000 | $230,000 | $137,000 | $480,000 |
| Horsepower | 485 bhp | 436 bhp | 375 bhp | 513 bhp |
| Weight | 3,725 lbs. | 4,325 lbs. | 3,390 lbs. | 2,710 lbs. |
| Engine Size | 334 cu. in. | 334 cu. in. | 213 cu. in. | 287 cu. in. |
| Top Speed | 199 mph | 185 mph | 175 mph | 202 mph |
| Miles per Gallon | 15 | 14 | 16 | 13 |
| Fuel Capacity | 30 gal. | 29.1 gal. | 21.6 gal. | 27.7 gal. |

1. What is the weight-to-power ratio of the 456 GTA?

2. Which car has the highest weight-to-power ratio? What is it?

3. Which car had the fastest time from a standing start to a quarter mile? What was its approximate time?

4. Express as a percent how much faster the F50 reaches 60 mph than the 550 Maranello reaches 60 mph.

5. Imagine you are racing the F50 in a 500-mile race. How many times will you have to stop to refuel?

6. Write the horsepower-to-engine displacement ratio of the F355 Spyder. Which car has the highest such ratio?

### Challenge Problem

Just for the fun of it, calculate the annual cost, over 5 years, of owning the Ferrari F355 Spyder. Assume that you will finance 90% of the car's suggested list price for 5 years at an annual percent of 11%. Call an insurance company and ask what the annual premium is for insuring the car. Assume that you will drive the car 12,000 miles each year. Figure out how much gas would cost each year based on the car's miles-per-gallon figure and the current cost of premium-grade gas. The manufacturer recommends an oil and filter change every 7,500 miles and a tune-up every 15,000 miles. Ask at a service station for the approximate costs of servicing the car. What is the total annual cost? Do you think any car is worth this much money? Why?

# Answer key

**Road race**
1. 56
2. 49
3. 47
4. 36
5. 24

**Magazine madness**
1. $46.89
2. $55.95
3. $40.89
4. $61.95
5. $130.84

**Speeding up**
1. 3:54
2. 128 mph
3. 3:05
4. 161 mph
5. 186 mph

**Autocross ladies**
1. 4.8 seconds
2. 47.8 seconds
3. 50.9 seconds
4. Kelly
Challenge problem: 50.15 seconds

**On the road**
1. 568 miles
2. 435 miles
3. 393 miles
4. 483 miles
Challenge problem: 3,250

**Winston Cup race**
1. 250 miles
2. 80 mph
3. More than 2 laps
4. $175,000
5. $58,000

**Fast laps**
1. 50 miles
2. 80 laps
3. 150 mph
4. 200 mph
5. Answers will vary.

**Favorite numbers**
Latisha, 80
Andre, 7
Patrick, 22
Jasmine, 13

Steffi, 72
Joel, 1
Roxanne, 54
Bart, 18

**Points to win**
1. 10
2. 10.2
3. 181; 162
4. de Ferran
5. 184; 150

**Drag racing**
1. 55 races
2. About 66%
3. $1,020
4. $1,836
Challenge problem: 42 races

**Day at the racetrack**
1. What was the total cost of their tickets?
2. What was the total cost of the tickets for the 8 kids or what was the total cost of the tickets for the 2 adults?
3. How many more grandstand seats than hillside seats were there?
4. How much would the 2 adults and 8 kids have paid if they all bought hillside seats?
5. How many of the seats were sold to adults?

## One-handed driving
1. 3-1/8 miles
2. 187.5 miles
3. 53.4 mph
4. 64.6 mph
5. 2.3 big wins per year

## Odd ovals

| | | |
|---|---|---|
| Hector | 310 miles | 91 mph |
| Courtney | 300 miles | 106 mph |
| Jermaine | 375 miles | 114 mph |
| Billy | 200 miles | 142 mph |

Challenge problem: Answers will vary. Generally, larger tracks have longer straightaways, resulting in faster speeds.

## Model money
1. $212.50
2. 10,500
3. $446,250
4. $29.25
5. He received $2,125. His profit was $1,912.50.
Challenge problem: 168 inches

## Indy 500 vs. U.S. 500
1. 0.5 mile
2. 8.447 mph
3. 10.3 seconds
4. 7.283 seconds
5. 4.961 mph
Challenge problem: Answers will vary.

## Hare racing
1. $186
2. $577.50
3. $35
4. $240
5. $519.60
Challenge problem: $1,520

## Pikes Peak hill climb
1. 9.58 seconds
2. 6.21 miles
3. 5:85.47
4. 1:43.16
5. 3:52.08

## Way graphic
1. $126
2. $441
3. $7,938
4. $121.88
5. $1,462.50
6. $313.35

## Racing suits and safety

| | | |
|---|---|---|
| Red | 7 seconds | $59.29 |
| Blue | 9.5 seconds | $55.68 |
| Green | 30 seconds | $26.27 |
| Yellow | 60 seconds | $22.37 |

## And laundry too
1. £40,292
2. £41,690
3. £20,480
4. £102,462
5. £19,635
Challenge problem: £1,439 = $2,302.40
£1,895 = $3,032.00
£10,240 = $16,384.00

## Gross tongues
1. 279 pounds
2. 418.5 pounds
3. 199.3 pounds
4. 298.95, or 299 pounds
5. 701.7 pounds

## Long day
1. 11.5 hours
2. 9 hours
3. $214.48
4. $643.44
5. $1,613.12

RacingMath

### Speedway capacity
1. $1,600,000
2. $24.60 per ticket
3. $9.80 per ticket
4. $14.70 per ticket

Challenge problem: More; $7,500 more

### Service with a smile
1. Windshield             180:5
   Brake pads            120:8
   Springs and shocks   360:15
   Engine                840:75
2. Windshield = 36
   Brake pads = 15
   Springs and shocks = 24
   Engine = 11.2
3. Windshield = 9
   Brake pads = 3.75
   Springs and shocks = 6
   Engine = 2.8

Challenge problem: Most efficient at changing the engine. Answers will vary.

### Points, points, points
1. Emerson Fittipaldi
2. Emerson Fittipaldi
3. Al Unser, Jr., and Robby Gordon
4. Al Unser, Jr.
5. Paul Tracy
6. No
7. No
8. Yes/Probably
9. Al Unser, Jr., probably had more first-place wins, or more first- and second-place wins than did Fittipaldi.

### Staggering around
1. Martinez 3/4"
   Carter 1-1/2"
   Dankovich 1-1/2"
   Tweet 1-3/4"
   Peebles 1-1/2"
2. Dankovich
3. Tweet

Challenge problem: Answers will vary, but in general a driver will want more stagger for a very tight oval track.

### Michael Andretti vs. Al Unser, Jr.
1. Andretti = 178 Starts, 30 Wins, 30 Poles
   Unser, Jr. = 204 Starts, 31 Wins, 7 Poles
2. Andretti = 30:30
   Unser, Jr. = 7:31
3. Andretti = 30:178
   Unser, Jr. = 31:204
4. Andretti = 17%
   Unser, Jr. = 15%
5. Andretti = 12
   Unser, Jr. = 14
6. Andretti = 2.5
   Unser, Jr. = 2.2

Challenge problem: Answers will vary.

### Metric chassis
1. 190 inches
2. 195 inches
3. 78.5 inches
4. 96 inches
5. 34.58 gallons

### First wins
1. 35
2. 18; 17; 17
3. 28
4. 1197; 34.2

Challenge problem: Answers will vary.

### Have a seat
1. 99.3 inches
2. Street; 442.9 square inches
3. Competition; 624.96 square inches
4. Competition; 21,123.65 cubic inches

Challenge problem: 0.018 oz. per cubic inch

### French Grand Prix
1. 22
2. 12
3. 54.5%
4. 4.5%
5. 4.5%
6. 18.2%
7. 9%
8. 33.3%
9. 16.6%
10. 27.2%

### Road rally
| | | | | |
|---|---|---|---|---|
| Team A | 50 | 54 | 0.96 | 57:36 |
| Team B | 49 | 51.58 | 0.95 | 57 |
| Team C | 50 | 57.47 | 0.87 | 52 |

4. Team B failed to cover the distance.
5. Team C failed to observe the speed limit.

### Big winnings
1. $11,986,391
2. $103,069,425
3. $41,070,005
4. $61,999,420

Challenge problem:
   Average = $7,206,971.50
   Median = $5,315,586.50

### Spider, spider
1. 180 bhp
2. 6,498 rmp
3. 1873.7 pounds
4. 96.05 mph
5. 2.655 times faster

### Television ratings
1. 
| | | |
|---|---|---|
| ABC | 1.7 | 1.6m |
| ABC | 2.2 | 2.1m |
| ABC | 1.6 | 1.5m |
| ABC | 6.6 | 6.3m |

2. 
| | |
|---|---|
| TBS | 673,913 |
| ESPN | 678,571 |
| TNN | 641,026 |

### Stock car growth
1. 1,900,000
2. 1.15%
3. 49.5%
4. 1,666,667
5. 15 million
6. $120
7. No; unlikely
8. Answers will vary; people who don't attend races probably buy merchandise.
9. $372,000,000

Challenge problem: 13.33%

### Not a drag
1. 34%
2. 8.5%
3. 10.89 seconds
4. 1\05.36%
5. 24.55%

### Victory's Vintage Storage
Starter
| | |
|---|---|
| Width | 50' |
| Depth | 20' |
| Area | 1,000 sq ft |
| Yearly | $28,000 |
| Monthly | $2,388.33 |

Pit stop
| | |
|---|---|
| Width | 60' |
| Depth | 22' |
| Area | 1,320 sq ft |
| Yearly | $36,960 |
| Monthly | $3,135 |

Final lap
| | |
|---|---|
| Width | 65' |
| Depth | 25' |
| Area | 1,625 sq ft |
| Yearly | $45,500 |
| Monthly | 3,846.67 |

RacingMath

Checkered flag
Width      70'
Depth      32'
Area       2,240 sq ft
Yearly     $62,720
Monthly    $5,281.67

**Tight turns**
1. 90°
2. 90°
3. 45°
4. 90°
5. 180°
6. 135°
7. 90°

**Home Sweet track**
1. $255,000; 268%
2. 24.4%
3. $1,204,000
4. $777,000
Challenge problem: $292,880; $1,342.37

**Michigan International Speedway**
1. 30.73 seconds
2. Rahal = 1:22:33
   Unser, Jr. = 2:38:07
3. 2:08.03
4. 37.95 seconds
5. 32.35 seconds

**Track size**
1. 0.35 mile
2. 0.7 mile
3. 0.5 mile
4. 1 mile
5. 1.7 miles

**Legends cars**
1. 96"
2. 201.6"
3. 73.6"
4. 1,880 lbs.
5. 200 hp
6. 20.8"
Challenge problem: 0.64 mile